Physics Research and Technology

nova
science publishers
www.novapublishers.com

Physics Research and Technology

Spatial Homeostasis, Quantum Information Channel, and the Nature of Living Things Within the Framework of the Theory of Byuon
Yuriy Baurov, PhD
2022. ISBN: 978-1-68507-590-3 (eBook)

A Closer Look at Plasmonics
Christian Bosch (Editor)
2022. ISBN: 978-1-68507-709-9 (Hardcover)
2022. ISBN: 978-1-68507-725-9 (eBook)

An Introduction to Charge Carriers
Jai Singh, PhD (Editor)
2022. ISBN: 978-1-68507-456-2 (Hardcover)
2022. ISBN: 978-1-68507-531-6 (eBook)

Handbook of Research on Heat Transfer
Suvanjan Bhattacharyya, PhD (Editor)
Varun Goel, PhD (Editor)
2022. ISBN: 978-1-68507-459-3 (Hardcover)
2021. ISBN: 978-1-68507-542-2 (eBook)

Solar PV System: Design, Installation, Operation and Maintenance
L. Ashok Kumar, PhD (Author)
K. Mohana Sundaram (Author)
2021. ISBN: 978-1-68507-341-1 (Hardcover)
2021. ISBN: 978-1-68507-475-3 (eBook)

More information about this series can be found at
https://novapublishers.com/product-category/newly-published-books/physics-nanotechnology/

Abdel-Baset M. A. Ibrahim and
Pankaj Kumar Choudhury
Editors

Second Harmonic Generation

Pathways of Nonlinear Photonics

nova
science publishers
www.novapublishers.com

DOI: https://doi.org/10.52305/CVTQ2987

NOTICE TO THE READER

The Publisher has taken reasonable care in the preparation of this book, but makes no expressed or implied warranty of any kind and assumes no responsibility for any errors or omissions. No liability is assumed for incidental or consequential damages in connection with or arising out of information contained in this book. The Publisher shall not be liable for any special, consequential, or exemplary damages resulting, in whole or in part, from the readers' use of, or reliance upon, this material. Any parts of this book based on government reports are so indicated and copyright is claimed for those parts to the extent applicable to compilations of such works.

Independent verification should be sought for any data, advice or recommendations contained in this book. In addition, no responsibility is assumed by the Publisher for any injury and/or damage to persons or property arising from any methods, products, instructions, ideas or otherwise contained in this publication.

This publication is designed to provide accurate and authoritative information with regard to the subject matter covered herein. It is sold with the clear understanding that the Publisher is not engaged in rendering legal or any other professional services. If legal or any other expert assistance is required, the services of a competent person should be sought. FROM A DECLARATION OF PARTICIPANTS JOINTLY ADOPTED BY A COMMITTEE OF THE AMERICAN BAR ASSOCIATION AND A COMMITTEE OF PUBLISHERS.

Additional color graphics may be available in the e-book version of this book.

Library of Congress Cataloging-in-Publication Data

ISBN: 978-1-68507-888-1

Published by Nova Science Publishers, Inc. † New York

To
All lost lives in the Covid pandemic

Contents

Preface ix

Chapter 1 **Visualizing the Scientific Landscape of Research in Second Harmonic Generation: A Scientometric Review** 1
Rafael Julius, Abdel-Baset M. A. Ibrahim and Pankaj K. Choudhury

Chapter 2 **Modeling Efficient Second Harmonic Generation in Microcrystalline KDP Fibers – Part I** 21
Matthew S. Neel

Chapter 3 **Modeling Efficient Second Harmonic Generation in Microcrystalline KDP Fibers – Part II** 49
Matthew S. Neel

Chapter 4 **Design and Analysis of Photonic Crystal-Based Integrated Optical Devices Using the Finite Difference Method** 83
Mohsen Choubani, Hassen Maaref and Faouzi Saidi

Chapter 5 **Nonlinear Optical Properties in Single and Coupled InAs/GaAs Quantum Dots** 117
Mohsen Choubani, Hassen Maaref and Faouzi Saidi

Chapter 6 **Quantum Properties of Light in Codirectional Coupler with Second Harmonic Generation** 151
Mohd Syafiq M. Hanapi, Abdel-Baset M. A. Ibrahim, Rafael Julius, Ahmad M. Alrub and Pankaj K. Choudhury

Chapter 7 **Quantum Dynamics of Contradirectional Nonlinear Coupler** 175
Mohd Syafiq M. Hanapi, Abdel-Baset M. A. Ibrahim, Rafael Julius, Ahmad M. Alrub and Pankaj K. Choudhury

Chapter 8 **Squeezed Light in Coupled Waveguide Networks Induced by the Nonlinear Kerr Effect and Second Harmonic Generation** 197
Rafael Julius, Abdel-Baset M. A. Ibrahim and Pankaj K. Choudhury

About the Editors 217

List of Contributors 219

Index 225

Preface

Second harmonic generation (SHG) has a wide range of applications in today's technological era, including nonlinear optics, quantum optics, lasers, material science, medical science, biological imaging, and high-resolution optical microscopy. In the laser industry, for example, SHG is prudent to create wavelength-specific high-energy lasers. It is also used to measure ultra-short pulse width with autocorrelators. SHG is now indispensable as a spectroscopic imaging tool in applications, such as biophysical characterization of the plasma membrane, biological sensing, disease diagnostics, and investigations of biomolecular interactions at interfaces. Because of its non-destructive detection, ultrafast response, and polarization sensitivity, SHG is exploited to describe crystal structures and materials. The use of SHG to characterize two-dimensional (2D) material structures gives crucial insights into their physical properties, thereby promoting the development of the relevant basic research, leading to the investigation of the potential applications of those materials. Developments in SHG research hold promising potentials of a large class of materials, such as magnetic- and nonmagnetic layered materials, perovskites, antiferromagnetic oxides, II-VI and III-V semiconductors, and nanotubes, for a variety of technological applications.

This book focuses on the process of modelling and simulations of the SHG phenomenon in the area of nonlinear and quantum optics. The first chapter provides a visualization of the scientific landscape of research in SHG using scientometric analysis from 1962 to 2020 based on Scopus database. This chapter gives new postgraduate students in the subject useful information on hot themes in SHG research and how they are related to one another. There is also a brief mention of multinational collaborative networks.

The following four research chapters look at the SHG from a classical standpoint, using Maxwell's equations to describe the nonlinear optical interaction between the electromagnetic wave and the medium. Such interaction is treated quantum mechanically in the second section of the book, with the SHG process described using a propagating Hamiltonian. As such, the volume adequately describes the SHG from both the classical and quantum mechanical standpoints. This allows the postgraduate researchers, focusing on the nonlinear phenomena, resulting from light-matter interaction, to find the content useful.

The theoretical foundation of the SHG coming from the interaction of an external field with a novel (monoclinic) polymorphic structure of potassium dihydrogen phosphate (KDP) crystal and its properties is covered in the first two chapters. Chapter 3 introduces the reader to a detailed finite difference time domain (FDTD) method for the analysis of several practical situations in the field of photonics. Chapter 4 uses the density matrix formalism, effective mass, and finite difference method (FDM) to investigate the linear and nonlinear optical properties (NLOP) in single quantum dot (SQD), InAs/GaAs coupled quantum dots (CQDs), and core/shell quantum dots (CQDs).

In the second part of this volume, the readers are introduced to a full theoretical analysis of the quantum features generated in certain optical devices, such as a two-waveguide device working under the SHG and coupled waveguide arrays with the combined second- and third-order nonlinear effects. To be more specific, this part discusses how SHG-enabled devices might be a useful source of nonclassical light. This section remains relevant for postgraduate students commencing their studies in quantum optics, where the nonclassical phenomena, such as squeezing and entanglement, require a solid understanding of the underlying techniques, namely the phase space and the analytical perturbative methods.

The editors are grateful to the contributors for sharing their most recent findings under this cover. All contributors' patience till the completion of this volume is much appreciated. The editors also like to express their gratitude to the Nova Science Publishers team for nominating them for this project. Our respective institutes, the Faculty of Applied Sciences of the Universiti Teknologi MARA (UiTM), and the Institute of Microengineering and Nanoelectronics of the Universiti Kebangsaan Malaysia, have also been very supportive while the work was in progress.

Abdel-Baset M. A. Ibrahim
Pankaj K. Choudhury

Chapter 1

Visualizing the Scientific Landscape of Research in Second Harmonic Generation: A Scientometric Review

Rafael Julius[1], Abdel-Baset M. A. Ibrahim[2,*] and Pankaj K. Choudhury[3]

[1]Faculty of Applied Sciences, Universiti Teknologi MARA (UiTM) Perak, Tapah Campus, Perak, Malaysia
[2]Faculty of Applied Sciences, Universiti Teknologi MARA (UiTM), Selangor, Malaysia
[3]Institute of Microengineering and Nanoelectronics, Universiti Kebangsaan Malaysia, Selangor, Malaysia

Abstract

This chapter provides the visualization of the scientific landscape of research in second harmonic generation (SHG) using scientometric analysis. From 1962 to 2020, a total of 23536 journal articles were retrieved from the Scopus database for analysis on the publication trend, the most refereed research papers, hot topics in SHG research, and how these topics are related to one another. A glimpse of international collaboration networks is also touched upon. According to the study, SHG research has sparked widespread interest showing significant growth, and is predicted to continue expanding in terms of the number of publications released each year. Research on crystal growth and structure, as well as the search and characterization of novel nonlinear optical materials containing organic molecules, are among the most active disciplines in SHG research, with collagen research being one of the most well-known topics.

* Corresponding Author's Email: abdelbaset@uitm.edu.my.

In: Second Harmonic Generation
Editors: Abdel-Baset M. A. Ibrahim and Pankaj Kumar Choudhury
ISBN: 978-1-68507-888-1

Keywords: nonlinear optics, second harmonic generation, second-order nonlinearity

1. Introduction

Nonlinear optics refer to the study of various nonlinear phenomena that emerged as a consequence of the presence of very high light intensities in nonlinear media [1]. While the nonlinear properties of light have been hypothesized since 1930 [2], the first experimental proof was only discovered in the 1960s [3], prior to the development of lasers that enable access to high-intensity light [4]. Since then, nonlinear optics has become a rapidly growing scientific field in recent decades [5]. Developments in the field of nonlinear optics hold promise for a variety of important applications in modern technology, as the physical science of light is critical to enabling future technologies. Among the most celebrated topics in the vast field of nonlinear optics include research related to solitons [6–8], optical fibers [9–11], photons [12–14], and laser pulses [15–17]. In this context, development and analysis on various potential applications of the lowest-order nonlinear response, i.e., the second harmonic generation (SHG), have occupied much of the theoretical and experimental efforts. SHG is a nonlinear optical process that results in the generation of photons with doubled frequency, and thus, doubles the energy of the initial photon before contact with the nonlinear media [18]. This process is only possible with materials that have a non-zero second harmonic coefficient. Because of that, only non-centrosymmetric materials can emit SHG light. Some typical SHG materials include ammonium dihydrogen phosphate (ADP), B-barium borate (BBO), gallium selenide (GaSe), lithium iodate ($LiIO_3$), and lithium niobate ($LiNBO_3$).

There have recently been a variety of insightful reviews on research related to the SHG process, focusing on a recent state of progress as well as issues, difficulties, and future study courses in various areas of application. For example, in the form of non- centrosymmetric metals [19, 20], semiconductors [21], and plasmonic nanostructures [22]. Nonetheless, the majority of the reviews relevant to SHG are focused on particular subjects, and in a broader sense where SHG, in general, is concerned, fundamental questions, such as what is the nature, hidden research trends, and the direction of growth of SHG research have remained the subject of scholarly discussions as these are usually not covered in a systematic literature review. From a scientometric viewpoint [23], this analysis aims to establish a global mapping of research related to SHG. This approach allows for the mapping of SHG research patterns, the identification of the most influential research articles and most celebrated topics, how these topics are relevant to one another through scientometric clustering, and the potential for international network collaborations.

2. Methodology

The literature for this analysis was collected from the Scopus database in May 2021, with "second harmonic generation" as the central theme. On the advance search of Scopus, the following query string was used: TITLE-ABS-KEY ("second harmonic generation" OR "second-harmonic generation") AND (EXCLUDE(PUBYEAR,2021)) AND (LIMIT-TO(DOCTYPE, "ar")) and a total of 23,536 papers were collected from 1962 to 2020. By specifying the title, abstract, or keywords, the query string (TITLE-ABS-KEY) assists the search engine in capturing articles related to the central theme. Without the (DOCTYPE, "ar"), the search string returns an additional 8802 publications, which include review papers, conference papers, book chapters, erratum, letters, and other items. Since we want to concentrate on journal papers in this report, we used the query string LIMIT-TO (SRCTYPE, "j") to narrow the search results down to 22,936. These documents were analyzed and presented in the form of a graphical plot and tables to find the publication pattern, the most prominent publication, and the research focus.

Furthermore, in order to concentrate on recent developments, search results prior to 1999 were omitted, leaving a total of 15,024 publications from 2000 to 2020 for review using similarity software visualization, i.e., VOS viewer. For visualizing a scientific landscape, the VOS viewer, in particular, is extremely useful [24]. It also allows for the development, visualization, and review of other scientometric networks, such as those of research authors, journals, organizations, and individual publications. Keywords, in particular, are represented in these visualizations by a frame and a link. The frame size corresponded to the number of occurrences, while the link size reflects how involved the areas are concerning other topics. The greater the frames and links of a keyword, the more active the research area, and vice versa. Furthermore, the frequency of a field relationship is determined by the distance between frames. In general, a shorter distance means a more active relationship. In addition, international network collaboration was examined in the same way, while data on intra-national collaboration were gathered using Single Country Publication (SCP) data with the field code AFFILCOUNTRY. These files are available online for more detailed mapping. Figure 1 depicts a flowchart of data collection, search, and analysis strategy.

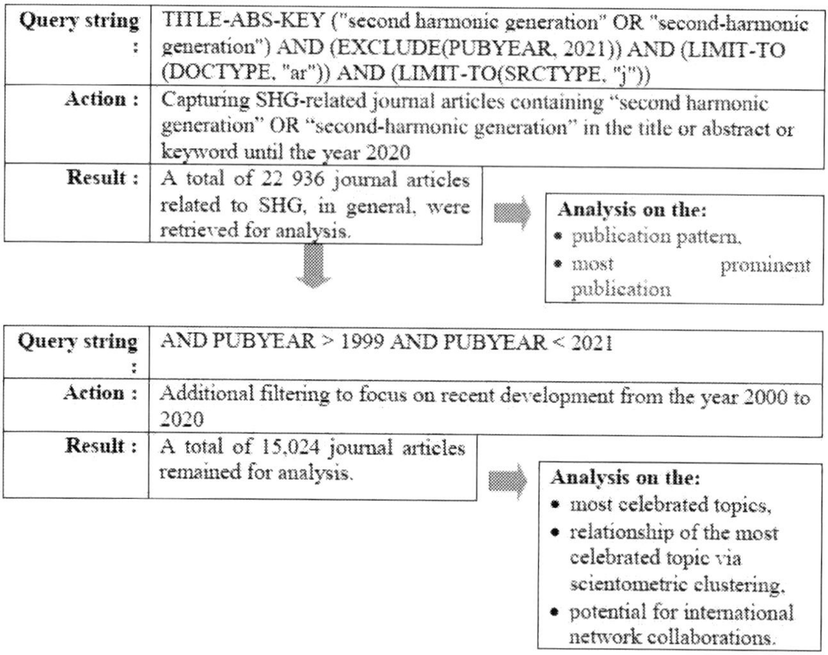

Figure 1. Flowchart of data collection, search and analysis strategy.

3. Discussion

3.1. The Growth Rate of SHG Research Publications

Figure 2 depicts the publication output on SHG research over 60 years from 1961 to 2020. The clustered column in the primary axis represents the annual growth of publication, while the solid line in the secondary axis represents the cumulative publication. This research focuses on 22 subject areas related to the core theme, with the most popular areas encompassing the field of SHG related to physics and astronomy, materials science, engineering, chemistry, and mathematics. The publication output for the first 26 years of the chosen era (1962–1986) is relatively low, with less than 100 publications published annually, suggesting that this research area was still in its early stages. R. L. Jepsen's measurements of SHG and second-order frequency mixing in ferromagnetic insulator [25] and D. D. Douthett, I. Kaufman, and A. S. Risley's experimental analysis of isotropic ferrimagnetic insulators in SHG at microwave frequencies [26] are among the earliest publications in the field of SHG research indexed in the Scopus database, dating back to 1961. Other than general keywords like harmonic generation and nonlinear optical effects, the research focus in this era remained on crystal studies and lasers, with an emphasis on the characterization of optical properties [27–29], synthesis [30, 31], irradiation [32, 33], thermal effects [34, 35] and others. In this era, the research also focused on the properties of ferroelectric materials, such as $LiNbO_3$ [36], $Pb_5Ge_3O_{11}$ [37], $SrTiO_3$ [38], $Cd(No_3)_2$ [39] and $Ba_2LiNb_5O_{15}$ [40], as well as the investigation and application of various types of lasers, such as dye [41–43], gas [44, 45], solid-state [46–48] and metal vapor [49] lasers.

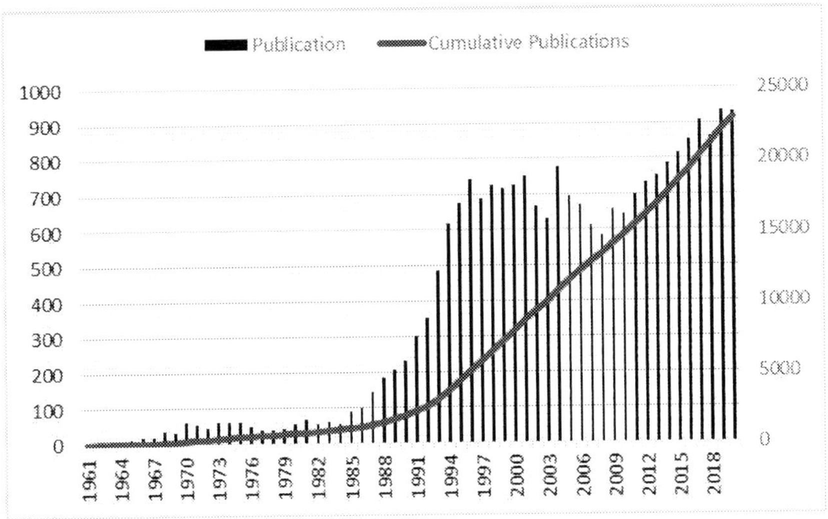

Figure 2. Distribution of publication on SHG research in Scopus.

A significant increase in the number of publications has been observed since 1987, indicating the growth of research interest. This may also mean a variety of other things, such as an increase in SHG research due to more publishing opportunities, more active researchers, increased research funding allocation, and expanded networks of collaborators. Interest in SHG research, with a focus on the application of lasers and crystals, remained strong after the year 1987. During this time, there has also been an increase in interest in the generation of SHG in optical waveguides. In particular, nonlinear waveguides propose a method for achieving efficient frequency doubling at low power levels. Among the numerous research papers published during this period, the most frequently cited research papers were those focusing on the generation of SHG in- channel waveguides with grating [50] and planar $LiNbO_3$ waveguides [51–53]. SHG research received a lot of attention from 1994 to 2006, although there was a small decrease in the number of publications from 2005 to 2008. SHG study via theoretical approach appears to be established at this point when studies related to mathematical models are frequently conducted with numerical calculations [54–56] and nonlinear equations [57–59]. The number of publications over the next 12 years (2009–2020) has reached a high level, with an average of 798 papers published per year, indicating that the field of SHG research has attracted widespread interest and reached a major growth and breakthrough. Based on this upward trend, it is expected that the number of annual publications will continue to rise, and SHG research will expand.

3.2. The Most Refereed Articles in SHG Research

Table 1 lists frequently cited articles with over 1000 citation counts in SHG research. M. Moskovits' paper *Surface-Enhanced Spectroscopy* [60] is one of the most widely cited papers on SHG. Providing an overview of surface-enhanced Raman spectroscopy, the article has been cited 4,839 times, with over 4,000 other publications referencing it. This is closely followed by the paper released by S. K. Kurtz and T. T. Perry, namely *A Powder Technique for the Evaluation of Nonlinear Optical Materials* [61], which has received 4,579 citations. The authors described an easy and fast experimental technique that allows for the rapid classification of materials based on the magnitude of nonlinear optical coefficients relative to a crystalline quartz standard and the presence or absence of phase-matching directions for SHG. Both of these papers are reviews published in authoritative journals, and despite being published in the early era where SHG research was still in its early stages, the articles are still important and widely cited in recent years. Furthermore, M. M. Fejer et al.,'s paper *Quasi-Phase-Matched Second Harmonic Generation: Tuning and Tolerances* [62] has been cited 1,948 times. The theory of quasiphase-matched SHG is discussed in this paper in both the space and wave-vector mismatch domains. To evaluate the tuning properties and acceptance bandwidths for SHG in periodic structures, deviations from ideal quasiphase matching in periodicity, wavelength, angle of propagation, and temperature are investigated. In another development, D. A. Kleinman discussed the physical mechanisms that can produce second-order dielectric polarization based on a simple extension of the theory of dispersion in ionic crystals. Four distinct mechanisms related to anharmonicity, second-order moment, lattice Raman scattering, and higher frequency electronic processes are defined. The paper *Nonlinear Dielectric Polarisation in Optical Media* [63] has earned 1,771 citations over 58 years. Furthermore, A. Yariv's article *Coupled-Mode Theory for Guided-Wave Optics* [64], which addresses the issue of optical radiation transmission and interaction in dielectric waveguides, has been cited 1,547 times.

Another widely cited scientific work is the theoretical research by G. D. Boyd and D. A. Kleinman on the optimization of SHG and parametric generation by a laser beam in a uniaxial nonlinear crystal. The paper *Parametric Interaction of Focused Gaussian Light Beams* [65] earned 1,456 citations. In this review, D. A. Kleinman is the only author with two papers earning over 1000 citations in SHG research. In their experimental analysis of an optically perfect single crystal, C. Chen et al., identified the boron-oxygen compound LiB_3O_5 as a new nonlinear-optical crystal. In addition to this, theoretical calculations of SHG coefficients using anionic group theory and the complete neglect of differential overlap approximation to obtain localized wave functions of member groups are discussed. The paper *New Nonlinear-Optical Crystal:* LiB_3O_5 [66] has been cited 1420 times since its publication in 1989. The articles by M. Fiebig et al., *Observation of Coupled Magnetic and Electric Domains* [67] and W. R. Zipfel et al., *Live Tissue Intrinsic Emission Microscopy Using Multiphoton-Excited Native Fluorescence and Second Harmonic Generation* [68], both published in 2002 and 2003, are the most recent additions to the list. Nonetheless, their work on spatial maps of coupled antiferromagnetic and ferroelectric domains in $YMnO_3$ and multicolor nonlinear microscopy of living tissue has had a major effect on the field of SHG research, with both papers receiving 1,243 and 1,418 total citations, respectively. Furthermore, J. L. Oudar's paper *Optical Nonlinearities of Conjugated Molecules. Stilbene Derivatives and Strongly Polar Aromatic Compounds* [69] has 1,413 citations. This paper provides a thorough experimental investigation of the quadratic and cubic hyperpolarizabilities of stilbene derivatives with different substituents. Next on the list is the paper *Supramolecular Materials: Self-Organized Nanostructures* [70] by S. I. Stupp et al., which has been cited 1,007 times. This paper investigates techniques for producing a large number of supramolecular units of similar size and shape.

Table 1. List of the most cited articles in SHG research

	Title and Year of Publication	Author(s)	TC	Source
1	Surface-enhanced spectroscopy	M. Moskovits	4839	Reviews of Modern Physics
2	A powder technique for the evaluation of nonlinear optical materials	S. K. Kurtz and T. T. Perry	4579	Journal of Applied Physics
3	Quasi-Phase-Matched Second Harmonic Generation: Tuning and Tolerances	M. M. Fejer et al.,	1948	IEEE Journal of Quantum Electronics
4	Nonlinear dielectric polarization in optical media	D. A. Kleinman	1771	Physical Review
5	Coupled-Mode theory for guided-wave optics	A. Yariv	1547	IEEE Journal of Quantum Electronics
6	Parametric interaction of focused Gaussian light beams	G. D. Boyd, and D. A. Kleinman	1456	Journal of Applied Physics
7	New nonlinear-optical crystal: LiB_3O_5	Chen, C. et al.,	1420	Journal of the Optical Society of America B: Optical Physics
8	Live tissue intrinsic emission microscopy using multiphoton-excited native fluorescence and second harmonic generation	W. R. Zipfel et. al	1418	Proceedings of the National Academy of Sciences of the United States of America
9	Optical nonlinearities of conjugated molecules. Stilbene derivatives and highly polar aromatic compounds	J. L. Oudar	1413	The Journal of Chemical Physics
10	Observation of coupled magnetic and electric domains	M. Fiebig et al.,	1243	Nature
11	Supramolecular materials: Self-organized nanostructures	S. I. Stupp et al.,	1007	Science

TC: Total Citations.

3.3. Hot Topics in SHG Research

To identify the research focus in SHG-related publications, occurrences of author keywords were analyzed, for which the most celebrated topic in SHG research has been recognized. Following the relabeling of keywords with minor differences, 13,937 keywords remained for analysis. The minimum number of keyword occurrences has been set to at least 30 times to focus only on the most popular topics, and 77 out of 13,937 keywords meet the threshold. In other words, all 77 keywords must appear on at least 30 documents. It should be noted that this analysis focused on recent developments from 2000 to 2020, and that additional filtering was applied to the main search results, resulting in a total of 15,024 journal articles. Figure 3 depicts a scientometric map of the author's keyword co-occurrence in network visualization mode, while Table 3 lists the most celebrated topic or keywords used to represent the field. These keywords were analyzed and arranged in clusters as shown in Table 4. Certainly, the keywords 'second harmonic generation' and 'nonlinear optics' received more attention than other keywords, with total occurrences of 3,048 and 918, respectively. Other popular alternative keywords in representing the field include the acronym 'SHG' and 'second-harmonic generation.'

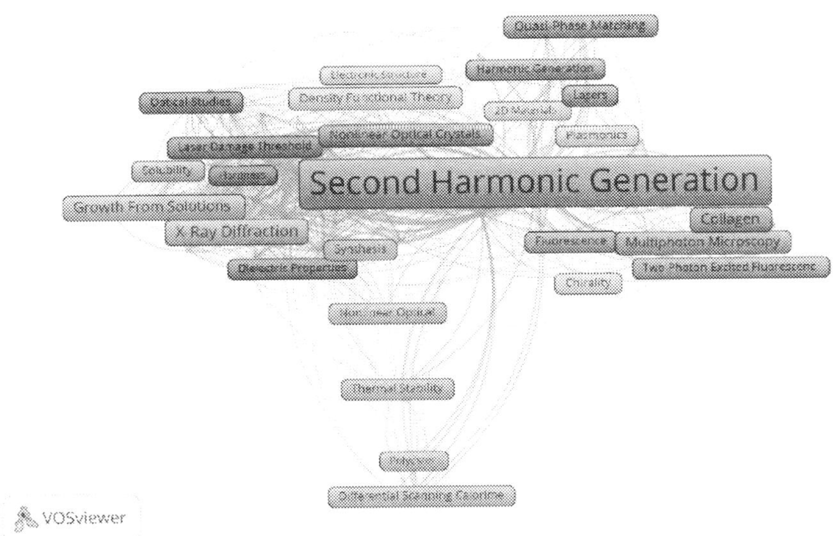

Figure 3. Scientometric map of author keywords co-occurrence. VOS viewer file is available at https://bit.ly/3g8NiMy.

Research related to crystal growth and structure is the most active field in SHG research, among many others. In this analysis, the keywords 'crystal growth,' 'crystal structure,' and 'growth from solution' have the highest frequency score, with 469, 207, and 215 occurrences, respectively, and 870, 276, 486 total links to other keywords. Aside from crystal studies, a large amount of effort in SHG research is devoted to the search or characterization of novel nonlinear optical materials involving organic compounds, with collagen research being one of the most well-known topics. This is demonstrated by the identification of keywords such as 'nonlinear optical materials' (427 occurrences, 763 total link), 'optical properties' (223 occurrences, 386 total link), 'optical materials' (102 occurrences, 244 total link), 'characterization' (119 occurrences, 242 total link), 'collagen' (237 occurrences, 319 total link), and 'organic compound' (98 occurrences, 264 total link). When compared to all other keywords, the keyword 'collagen' has the highest link strength of 134 to the keyword 'second harmonic generation.' Furthermore, other most popular keywords, as shown in Table 3, are related to the technique or methodology associated with the generation SHG, characterization, and imaging, such as X-Ray Diffraction (XRD) (406 occurrences, 937 total link), Fourier-Transform Infrared (FTIR) (134 occurrences, 277 total link), and Multiphoton Microscopy (154 occurrences, 196 total link). Besides, among the many types of analyses involved, the keyword 'thermal analysis' appears the most (82 times) with 206 total link strength.

Table 3. List of the most celebrated topic in SHG research based on the total link strength

Keywords	Total Link Strength
Second Harmonic Generation	2028
Nonlinear Optics	1024
Crystal Growth	870
X-Ray Diffraction (XRD)	937
Nonlinear Optical Materials	763
Growth from Solutions	486
Optical Properties	386
Collagen	319
FTIR	277
Crystal Structure	276
Organic Compounds	264
Optical Materials	244
Characterization	242
Thermal Analysis	206
Multiphoton Microscopy	196

Clustering techniques play an important role in visualizing the scientific landscape of research; for example, in the current review, it can be used to recognize many related publications. This illustration depicts the most prominent features or keywords that appear in publications as well as the co-occurrence relationships between them. In Table 4, the scientometric clustering of the most popular topic in SHG research is given where all 77 keywords were organized into 7 different clusters. The biggest cluster belongs to the research associated with crystal studies. This indicates that research in this particular area has produced most of the papers in the field of SHG research. There is also evidence of crystal studies in another cluster, for example, keywords 'single crystal growth' and 'crystal structure' in Cluster 5 and Cluster 7, respectively. Nonetheless, based on the citation relations, these keywords are more closely related to the other topic of interest, and are thus clustered in different groups. Because crystals are actively studied in SHG research, keywords like 'crystal growth,' 'single crystal,' and 'solution growth' are frequently associated with optical and dielectric characterizations using prevalent approaches like FTIR [71–73] and XRD [74–76]. In addition, Cluster 1 covered, among other things, research on the mechanical and thermal properties of optical materials as well.

Furthermore, the second-largest cluster appears to be related to SHG research involving a three-dimensional network consisting of extracellular macromolecules and minerals, such as collagen [77–79], which is frequently associated with keywords referring to the use of various imaging techniques, such as microscopy- and fluorescence- based approaches. Cluster 3 appeared to be made up of general terms representing the study of nonlinear optics, specifically SHG. Because 'nonlinear optics' is the most general keyword that represents the entire field, it is natural for it to be located around the center of the map that connects all the other keywords after 'second harmonic generation.' For this reason, Cluster 3 is frequently associated with keywords related to the description of obtaining SHG, such as 'harmonic generation' (72 occurrences, 48 total link), 'frequency conversion' (84 occurrences, 84 total link), 'quasi-phase matching' (101 occurrences, 105 total link), 'phase matching' (49 occurrences, 59 total link), 'frequency doubling' (34 occurrences, 25 total link) and 'conversion efficiency' (31 occurrences, 35 total link).

One of the most difficult aspects of advanced science experimentation is that it frequently leads to increased expenses. For this case, the theoretical model may provide significantly more flexibility. The Density Functional

Theory (DFT) is one of the most distinct conceptual modeling techniques that has been commonly used to study SHG. DFT is a validated computational quantum mechanical analysis tool that can be used to compute molecular and electronic structures. We discovered that the keyword 'Density Functional Theory' had 86 occurrences with 99 total link strength in Cluster 4. Furthermore, DFT is inextricably linked to the study of electronic structures, hyperpolarizability, and nonlinear optical properties [80–83]. At the far end of the cluster (Figure 3), DFT is also concerned with the work of 2D materials and plasmonic research. Other topics of study within the same cluster include nonlinear optical properties (such as SHG) and chirality in thin films and silicon. The hot topics of research in Cluster 5 are centered on nonlinear optical materials, with a strong emphasis on keywords, such as 'single crystal growth' (73 occurrences, 167 total links) and 'growth from solutions' (215 occurrences, 486 total links). Cluster 5 was also concerned with SHG study in organic compound and characterization, such as solubility using XRD as the main analysis tools which are very similar to the topics covered in Cluster 1, and hence, explained the proximity of the two clusters.

Table 4. Scientometric clustering of the most popular topic in SHG research

Cluster 1	Cluster 2	Cluster 3
Crystal Growth	Autofluorescence	Borates
Dielectric	Collagen	Conversion Efficiency
Dielectric Properties	Elastic	Frequency Conversion
Dielectric Studies	Extracellular Matrix	Frequency Doubling
FTIR	Femtosecond Laser	Harmonic Generation
Hardness	Fluorescence	Lasers
Laser Damage Threshold	Imaging	Lithium Niobate
Mechanical Properties	Microscopy	Nonlinear Optical Crystals
Microhardness	Multiphoton Microscopy	Nonlinear Optics
Optical Materials	Nonlinear Microscopy	Phase Matching
Optical Properties	Nonlinear Optical Microscopy	Quasi-Phase Matching
Optical Studies	Polarization	Cluster 4
Photoluminescence	Second Harmonic Generation	2D Materials
Raman Spectroscopy	Second Harmonic Generation Microscopy	Chirality
SHG Efficiency	Third Harmonic Generation	Density Functional Theory
Single Crystal	Two-Photon Excited Fluorescence	Electronic structure
Solution growth	Two-Photon Microscopy	Hyperpolarizability
Thermal Analysis		Nonlinear Optical Properties
Thermal Properties		Plasmonic
XRD		Silicon
Z-Scan		Thin Films

Cluster 5	Cluster 6	Cluster 7
Characterization Growth from Solutions Nonlinear Optical Materials Organic Compounds Single Crystal Growth Solubility X-Ray Diffraction	Differential Scanning Calorimetry (DSC) Nonlinear Optical Polyester SHG Coefficient Thermal Stability Thermogravimetric Analysis (TGA)	Crystal Structure Luminescence Nonlinear Phase Transition Synthesis

Another well-known research topic in this analysis is the calculation of thermal stability of SHG coefficients involving molecular design and synthesis of polymer chemistry [84–86] (Cluster 6). Nonetheless, upon further examination, we discovered that these studies are somewhat isolated from the rest of the clusters, implying that Cluster 6 stands on its own, and that the likelihood of keywords in this cluster co-occurring with keywords from the other clusters is less likely. Even though Cluster 7 contains the fewest keywords, the research topics covered in Cluster 7 are very dominant in the field, particularly in terms of the study of crystal structure and synthesis. As shown in Figure 3, the keyword 'crystal structure' is located around the center of the map, surrounded by a slew of other keywords. Because these clusters have been normalized using the association strength measure, it appears that research topics in Cluster 7 are highly relevant to other topics and are more likely to co-occur with other keywords in a large number of documents. Readers are directed to the online map, which can be accessed by using the VOS viewer at https://bit.ly/3g8NiMy, for more detailed information, such as the relationship level between each topic in the same cluster or between clusters, the timeline of the papers being published, or keyword co-occurrence and so on.

3.4. International Collaboration Network

A strong collaborative network across institutions or countries has shown efficacy not only in terms of adapting different concepts and ideas, but also in terms of broadening access to a broader range of methodological approaches and possibly accelerating scientific progress. In Figure 4, the visualization of international collaboration between selected countries with at least 20 publications is presented. Of the 137 countries extracted from 15,024 journal articles, 52 countries meet the thresholds. In this overlay visualization mode, countries with a significant number of co-authorship publications are positioned close to each other with thicker connecting lines, and vice versa, whereas the color scheme indicates the timeline for when the networks are being established. In addition, the size of the frame correlates to the significance of countries in SHG research. China has been identified as the largest contributor with 3,658 total publication (TP), accounting for 24.3% of total global publications, followed by the United States (US) (2,863 TP; 19.0%), India (1,766 TP; 11.7%), Japan (1,452; 9.7%) and Germany (1,258; 8.4%).

The US and Germany are the most active international collaborators, each with ties to 48 countries. India and China, on the other hand, have a strong intracountry collaboration network, with only 16.6% and 23.3% of their total publication affiliated with other countries, respectively. In terms of the contributions of the educational universities in these countries, the majority of publications are from the Chinese Academy of Sciences (China), Northwestern University (US), Anna University (India), Tokyo Institute of Technology (Japan), and Friedrich Schiller University Jena (Germany). Furthermore, the US's occupation of the center map indicates that, among other nations, the US plays the most significant role in establishing connections between scientific communities. The highest number of co-authorship papers are recorded between the US and China (322 co-authorship), Germany (184 co-authorship), the United Kingdom (104 co-authorship), France (104 co-authorship), Italy (98 co-authorship).

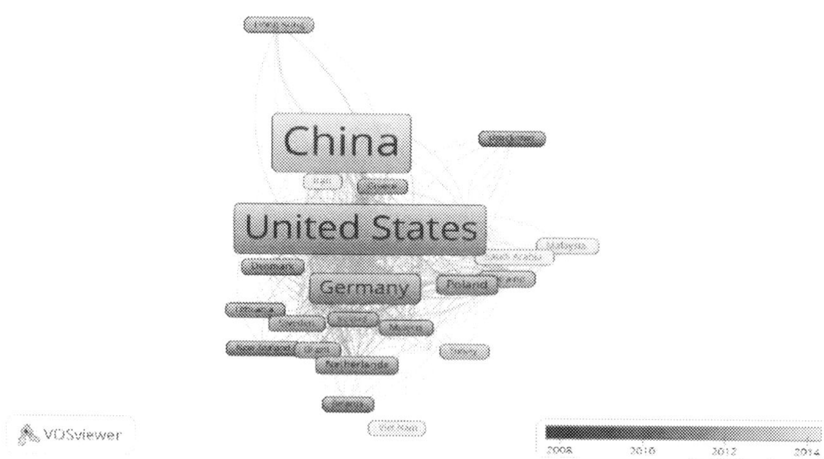

Figure 4. Scientometric map of international network collaboration. VOS viewer file is available at https://bit.ly/3g7m2hj.

The collaborative network depicted in Figure 4 has been organized into clusters based on the merits of co-authorship between countries. The VOS viewer can provide detailed information on the networking scheme. The reader is directed to the map at https://bit.ly/3g7m2hj for more detailed mapping.

Conclusion

In this chapter, a scientometric analysis of global research performance and analysis of SHG research was conducted with a total of 23,536 journal articles retrieved from the Scopus database from 1962 to 2020. According to the analysis, the field of SHG research has attracted widespread interest, experienced significant growth, and is expected to continue increasing in the number of publications each year. All of the 77 most prevalent keywords in SHG study have been categorized into 7 separate clusters to provide information on which keywords are more connected to than others. The keywords 'second harmonic generation' and 'nonlinear optics' describe the subject of research in general, and among many others, research related to crystal growth and structure are the most active fields in SHG research followed by research related to the search or characterization of novel nonlinear optical materials involving organic compounds, with collagen research being one of the most well-known topics.

Acknowledgments

R Julius gratefully recognizes the Malaysian Ministry of Higher Education (MOHE) national grant FRGS/1/2021/STG07/UITM/02/4.

References

[1] Zheltikov, A., A. L'Huillier, and F. Krausz, "Nonlinear Optics," in *Springer Handbook of Lasers and Optics*, F. Träger, Ed. New York, NY: Springer, pp. 157–248, 2007.

[2] Suresh, S., A. Ramanand, D. Jayaraman, and P. Mani, "Review on theoretical aspect of nonlinear optics," Rev. Adv. Mater. Sci., vol. 30, no. 2, 2012.

[3] Franken, P. A., A. E. Hill, C. W. Peters, and G. Weinreich, "Generation of optical harmonics," *Phys. Rev. Lett*., vol. 7, no. 4, pp. 118–119, 1961.

[4] Hecht, J. "Short history of laser development," *Opt. Eng*., vol. 49, no. 9, p. 091002, 2010.

[5] Kivshar, Y. S. "Nonlinear optics: the next decade," *Opt. Express*, vol. 16, no. 26, p. 22126, 2008.

[6] Kauranen, M. and A. V. Zayats, "Nonlinear plasmonics" *Nat. Photonics*, vol. 6, no. 11, pp.737-748, 2012.

[7] Fleischer, J. W., M. Segev, N. K. Efremidis, and D. N. Christodoulides, "Observation of two-dimensional discrete solitons in optically induced nonlinear photonic lattices," *Nature*, vol. 422, no. 6928, pp. 147-150, 2003.

[8] Musslimani, Z., K. Makris, R. El-Ganainy, and D. Christodoulides, "Optical solitons in PT periodic potentials," *Phys. Rev. Lett*., vol. 100, no. 3, p. 030402, 2008.

[9] Erdogan, T. "Fiber grating spectra," *J. Light. Technol*., vol. 15, no. 8, pp. 1277–1294, 1997.

[10] Russell, P. S. J. "Photonic-crystal fibers," *J. Light. Technol.*, vol. 24, no. 12, pp. 4729–4749, 2006.

[11] Stockman, M. I. "Nanofocusing of optical energy in tapered plasmonic waveguides," *Phys. Rev. Lett.*, vol. 93, no. 13, p. 137404, 2004.

[12] Akahane, Y., T. Asano, B.-S. Song, and S. Noda, "High- Q photonic nanocavity in a two-dimensional photonic crystal," *Nature,* vol. 425, no. 6961, p. 6961, 2003.

[13] Carusotto, I. and C. Ciuti, "Quantum fluids of light," *Rev. Mod. Phys*., vol. 85, no. 1, pp. 299–366, 2013.

[14] Duan, L. M. and H. J. Kimble, "Scalable photonic quantum computation through cavity-assisted interactions," *Phys. Rev. Lett.*, vol. 92, no. 12, p. 127902, 2004.

[15] Fleischhauer, M., A. Imamoglu, and J. P. Marangos, "Electromagnetically induced transparency: Optics in coherent media," *Rev. Mod. Phys.*, vol. 77, no. 2, pp. 633–673, 2005.

[16] Turukhin, A. V., V. S. Sudarshanam, M. S. Shahriar, J. A. Musser, B. S. Ham, and P. R. Hemmer, "Observation of ultraslow and stored light pulses in a solid," *Phys. Rev. Lett.*, vol. 88, no. 2, p. 023602, 2001.

[17] Schaffer, C., A. Brodeur, J. García, and E. Mazur, "Micromachining bulk glass by use of femtosecond laser pulses with nanojule energy," *Opt. Lett.*, vol. 26, pp. 93–5, 2001.

[18] Paschotta, R., article on 'Frequency Doubling' in the *Encyclopedia of Laser Physics and Technology*, 1st Ed. Wiley-VCH, ISBN 978-3-527-40828-3., 2008.

[19] Orenstein, J., J. E. Moore, T. Morimoto, D. H. Torchinsky, J. W. Harter, and D. Hsieh, "Topology and symmetry of quantum materials via nonlinear optical responses," *Annu. Rev. Condens. Matter Phys.*, vol. 12, no. 1, pp. 247–272, 2021.

[20] Zeng, Z. X. S., X. Wang, and A.-L. Pan, "Second harmonic generation of two-dimensional layered materials: characterization, signal modulation and enhancement," *Acta Phys. Sin.*, vol. 69, no. 18, p. 20200452, 2020.

[21] Gigli, C., G. Marino, A. Borne, P. Lalanne, and G. Leo, "All-dielectric nanoresonators for $\chi^{(2)}$ nonlinear optics," Front. Phys., vol. 7, p. 221, 2019.

[22] Butet, J., P. F. Brevet, and O. Martin, "Optical second harmonic generation in plasmonic nanostructures: from fundamental principles to advanced applications," *ACS Nano*, vol. 9, no. 11, pp. 10545-10562, 2015.

[23] van Raan, A. "Measuring Science: Basic Principles and Application of Advanced Bibliometrics," in *Springer Handbook of Science and Technology Indicators*, W. Glänzel, H. F. Moed, U. Schmoch, and M. Thelwall, Eds. Cham: Springer International Publishing, pp. 237–280, 2019.

[24] van Eck, N. J. and L. Waltman, "Software survey: VOSviewer, a computer program for bibliometric mapping," *Scientometrics*, vol. 84, no. 2, pp. 523–538, 2010.

[25] Jepsen, R. L., "Harmonic generation and frequency mixing in ferromagnetic insulators," *J. Appl. Phys.*, vol. 32, no. 12, pp. 2627–2630, 1961.

[26] Douthett, D., I. Kaufman, and A. Risley, "Microwave harmonic generation by ferrimagnetic crystals," Appl. Phys., vol. 32, no. 10, pp. 1905-1912, 1961.

[27] Byer, R. L. "Nonlinear optical phenomena in oxides and application to a tunable coherent spectrometer," *J. Solid State Chem.*, vol. 12, no. 3, pp. 156–171, 1975.

[28] Boyd, G., H. Kasper, J. McFee, F. Storz, "Linear and nonlinear optical properties of some ternary selenides," *IEEE J Quantum Electron*, vol. 8, no. 12, pp. 900–908, 1972.

[29] Iio, K. "Nonlinear optical property of sodium nitrite. I. second harmonic generation at room temperature," *J. Phys. Soc. Jpn.*, vol. 34, no. 1, pp. 138–147, 1973.

[30] Robbins, M. and V. G. Lambrecht Jr., "Preparation and some properties of materials in systems of the type MIMIIIS2MIMIIISe2 where MI = Cu, Ag and MIII = Al, Ga, In," *Mater. Res. Bull.*, vol. 8, no. 6, pp. 703–709, 1973.

[31] Demina, L. A., V. A. Dolgikh, S. Yu. Stefanovich, S. A. Okonenko, and B. A. Popovkin, "Synthesis and Certain Properties of beta -Sb2O4 Single Crystals.," *Neorganiceskie Mater.*, vol. 16, no. 3, pp. 470–474, 1980.

[32] Volosov, V. D., A. M. Dykhovnyi, V. N. Krylov, and T. V. Sokolova, "Second-harmonic conversion of laser radiation generated under free-oscillation conditions.," *Sov. J. Quantum Electron*, vol. 2, no. 2, pp. 175–176, 1972.

[33] Lokhov, Yu. N., V. S. Mospanov, and Yu. D. Fiveiskii, "Maximum permissible pulse duration in second-harmonic generation in A KDP crystal," *Sov. J. Quantum Electron*, vol. 2, no. 2, pp. 177–179, 1972.

[34] Pandarese, F., S. Panizza, and A. Telò, "Temperature effects on second harmonic generation in p-nitrophenylhydrazine," *Opt. Commun.*, vol. 12, no. 1, pp. 14–16, 1974.

[35] Dmitriev, V. G., V. A. Konovalov, and E. A. Shalaev, "Theory of thermal self-action in second-harmonic generation in nonlinear crystals," *Sov. J. Quantum Electron.*, vol. 5, no. 3, p. 282, 1975.

[36] Paraire, N., S. Laval, and M. Chapet-Rousseau, "Determination of some optical constants of LiNbO3 by backward stimulated Raman Scattering," *Appl. Phys.*, vol. 21, no. 2, pp. 115–120, 1980.

[37] Miller, R. C., W. A. Nordland, and A. A. Ballman, "Nonlinear optical properties of ferroelectric Pb5Ge3O11," *Opt. Commun.*, vol. 6, no. 2, pp. 210–212, 1972.

[38] Betzler, K. "Second harmonic generation near the phase transition of SrTiO3," *Ferroelectrics*, vol. 26, no. 1, pp. 819–822, 1980.

[39] Bernard, D. and D. Louer, "Phase transition in cadmium nitrate, Cd(NO3)2," *Ferroelectrics*, vol. 25, no. 1, pp. 461–463, 1980.

[40] Bernecker, O., H. Matthes, and A. Marshall, "Optical properties of Ba2LiNb5O15," *Phys. Status Solidi A*, vol. 17, no. 2, pp. 453–458, 1973.

[41] Saikan, S. "Automatically tunable second-harmonic generation of dye lasers," *Opt. Commun.*, vol. 18, no. 4, pp. 439–443, 1976.

[42] Ferguson, A. I. and M. H. Dunn, "Intracavity second-harmonic generation in continuous-wave dye lasers," *IEEE J. Quantum Electron.*, vol. 13, no. 9, pp. 751–756, 1977.

[43] Hirth, A., K. Vollrath, and J. Y. Allain, "Production of high power tunable UV laser emission by second-harmonic generation fro, a rhodamine 6G dye laser," *Opt. Commun.*, vol. 20, no. 3, pp. 347–349, 1977.

[44] Ng, A., M. R. Cervenan, and A. A. Offenberger, "A gas target for the study of laser-induced parametric instabilities in plasmas," *Opt. Commun.*, vol. 25, no. 1, pp. 116–120, 1978.

[45] Nikogosyan, D. N., A. P. Sukhorukov, and M. I. Golovei, "Saturation in generation of second harmonic of tea co2 laser radiation in proustite.," *Sov J Quantum Electron*, vol. 5, no. 3, pp. 344–346, 1975.

[46] Guenter, P., P. M. Asbeck, M. M. Choy, and S. K. Kurtz, "Second-harmonic generation with Ga1-xAlxAs lasers and KNbO3 crystals," *Ferroelectrics*, vol. 28, no. 1, pp. 379–382, 1980.

[47] Culsha, W., J. Kannelaud, and J. E. Peterson, "Efficient frequency-doubled single-frequency nd:yag laser," *IEEE J. Quantum Electron.*, vol. QE-10, no. 2, pp. 253–263, 1974.

[48] Uematsu, Y. and T. Fukuda, "Characteristics and performance of knbo3-yalg/nd intracavity second harmonic generation," *Jpn. J. Appl. Phys.*, vol. 12, no. 6, pp. 841–844, 1973.

[49] Isaev, A. A., G. Yu. Lemmerman, and G. L. Malafeeva, "Second harmonic generation from pulsed copper vapor laser radiation," *Sov. J. Quantum Electron.*, vol. 10, no. 8, pp. 983–985, 1980.

[50] Farries, M. C., P. S. J. Russell, M. E. Fermann, and D. N. Payne, "Second-harmonic generation in an optical fibre by self-written $x^{(2)}$ grating," *Electron. Lett.*, vol. 23, no. 7, pp. 322–324, 1987.

[51] Lim, E. J., M. M. Fejer, and R. L. Byer, "Second-harmonic generation of green light in periodically poled planar lithium niobate waveguide," *Electron. Lett.*, vol. 25, no. 3, pp. 174–175, 1989.

[52] Suhara, T., H. Nishihara, and H. Tazaki, "Measurement of reduction in SHG coefficient of LiNb03 by proton exchanging," *Electron. Lett.*, vol. 25, no. 20, pp. 1326–1328, 1989.

[53] Luh, Y. S., M. M. Fejer, R. L. Byer, and R. S. Feigelson, "Stoichiometric LiNbO3 single-crystal fibers for nonlinear optical applications," *J. Cryst. Growth*, vol. 85, no. 1–2, pp. 264–269, 1987.

[54] Kador, L. "Kramers-Kronig relations in nonlinear optics," *Appl. Phys. Lett.*, vol. 66, p. 2938, 1995.

[55] Masoudi, H. M. and J. M. Arnold, "Parallel beam propagation method for the analysis of second harmonic generation," *IEEE Photonics Technol. Lett.*, vol. 7, no. 4, pp. 400–402, 1995.

[56] Popov, E. and M. Nevière, "Surface-enhanced second-harmonic generation in nonlinear corrugated dielectrics: New theoretical approaches," *J. Opt. Soc. Am. B Opt. Phys.*, vol. 11, no. 9, pp. 1555–1564, 1994.

[57] Hall, C. D. "Averaging of second-order hamiltonian oscillators with a slowly varying parameter," *J. Guid. Control Dyn.*, vol. 16, no. 4, pp. 782–784, 1993.

[58] Kioka, W. and H. Ishida, "Green's function approach to the weakly nonlinear interaction problem of two-dimensional structures of arbitrary cross-section," *Doboku Gakkai Rombun-Hokokushu*, no. 461 pt 2–22, pp. 59–65, 1993.

[59] Kalocsai, A. G. and J. W. Haus, "Nonlinear Schrödinger equation for optical media with quadratic nonlinearity," *Phys. Rev. A*, vol. 49, no. 1, pp. 574–585, 1994.

[60] Moskovits, M. "Surface-enhanced spectroscopy," *Rev. Mod. Phys.*, vol. 57, no. 3, pp. 783–826, 1985.

[61] Kurtz, S. K. and T. T. Perry, "A powder technique for the evaluation of nonlinear optical materials," *J. Appl. Phys.*, vol. 39, no. 8, pp. 3798–3813, 1968.

[62] Fejer, M. M., D. H. Jundt, R. L. Byer, and G. A. Magel, "Quasi-phase-matched second harmonic generation: tuning and tolerances," *IEEE J. Quantum Electron.*, vol. 28, no. 11, pp. 2631–2654, 1992.

[63] Kleinman, D. A. "Nonlinear dielectric polarization in optical media," *Phys. Rev.*, vol. 126, no. 6, pp. 1977–1979, 1962.

[64] Yariv, A. "Coupled-Mode theory for guided-wave optics," *IEEE J. Quantum Electron.*, vol. 9, no. 9, pp. 919–933, 1973.

[65] Boyd, G. D. and D. A. Kleinman, "Parametric interaction of focused Gaussian light beams," *J. Appl. Phys.*, vol. 39, no. 8, pp. 3597–3639, 1968.

[66] Chen, C. et al., "New nonlinear-optical crystal: LiB3O5," *J. Opt. Soc. Am. B Opt. Phys.*, vol. 6, no. 4, pp. 616–621, 1989.

[67] Zipfel, W. R., R. M. Williams, R. Christiet, A. Y. Nikitin, B. T. Hyman, and W. W. Webb, "Live tissue intrinsic emission microscopy using multiphoton-excited native fluorescence and second harmonic generation," *Proc. Natl. Acad. Sci. U. S. A.*, vol. 100, no. 12, pp. 7075–7080, 2003.

[68] Oudar, J. L., "Optical nonlinearities of conjugated molecules. Stilbene derivatives and highly polar aromatic compounds," *J. Chem. Phys.*, vol. 67, no. 2, pp. 446–457, 1977.

[69] Fiebig, M., T. Lottermoser, D. Fröhlich, A. V. Goltsev, and R. V. Pisarev, "Observation of coupled magnetic and electric domains," *Nature*, vol. 419, no. 6909, pp. 818–820, 2002.

[70] Stupp, S. I. et al., "Supramolecular materials: Self-organized nanostructures," *Science*, vol. 276, no. 5311, pp. 384–389, 1997.

[71] Rajesh, N. P., V. Kannan, M. Ashok, K. Sivaji, P. Santhana Raghavan, and P. Ramasamy, "A new nonlinear optical semi-organic material: Cadmium thiourea acetate," *J. Cryst. Growth*, vol. 262, no. 1–4, pp. 561–566, 2004.

[72] Skauli, T. et al., "Improved dispersion relations for GaAs and applications to nonlinear optics," *J. Appl. Phys.*, vol. 94, no. 10, pp. 6447–6455, 2003.

[73] Parikh, K. D., D. J. Dave, B. B. Parekh, and M. J. Joshi, "Thermal, FT-IR and SHG efficiency studies of L-arginine doped KDP crystals," *Bull. Mater. Sci.*, vol. 30, no. 2, pp. 105–112, 2007.

[74] Fiebig, M., et al., "Determination of the magnetic symmetry of hexagonal manganites by second harmonic generation," *Phys. Rev. Lett.*, vol. 84, no. 24, pp. 5620–5623, 2000.

[75] Yamada, H. et al., "Engineered interface of magnetic oxides," *Science*, vol. 305, no. 5684, pp. 646–648, 2004.

[76] Sykora, R. E., M. O. Kang, P. S. Halasyamani, and T. E. Albrecht-Schmitt, "Structural modulation of molybdenyl iodate architectures by alkali metal cations in AMoO3(IO3) (A = K, RB, Cs): A facile route to new polar materials with large SHG responses," *J. Am. Chem. Soc.*, vol. 124, no. 9, pp. 1951–1957, 2002.

[77] Provenzano, P. P., K. W. Eliceiri, J. M. Campbell, D. R. Inman, J. G. White, and P. J. Keely, "Collagen reorganization at the tumor-stromal interface facilitates local invasion," *BMC Med.*, vol. 4, 2006.

[78] Ingman, W. V., J. Wyckoff, V. Gouon-Evans, J. Condeelis, and J. W. Pollard, "Macrophages promote collagen fibrillogenesis around terminal end buds of the developing mammary gland," *Dev. Dyn.*, vol. 235, no. 12, pp. 3222–3229, 2006.

[79] Cicchi, R. et al., "Scoring of collagen organization in healthy and diseased human dermis by multiphoton microscopy," *J. Biophotonics*, vol. 3, no. 1–2, pp. 34–43, 2010.

[80] Wu, H. et al., "Designing a deep-ultraviolet nonlinear optical material with a large second harmonic generation response," *J. Am. Chem. Soc.*, vol. 135, no. 11, pp. 4215–4218, 2013.

[81] Stoumpos, C. C. et al., "Hybrid germanium iodide perovskite semiconductors: Active lone pairs, structural distortions, direct and indirect energy gaps, and strong nonlinear optical properties," *J. Am. Chem. Soc.*, vol. 137, no. 21, pp. 6804–6819, 2015.

[82] Liu, F. et al., "Room-temperature ferroelectricity in CuInP 2 S 6 ultrathin flakes," *Nat. Commun.*, vol. 7, 2016.

[83] van Gisbergen, S. J. A., J. G. Snijders, and E. J. Baerends, "Time-dependent density functional results for the dynamic hyperpolarizability of C_60," *Phys. Rev. Lett.*, vol. 78, no. 16, pp. 3097–3100, 1997.

[84] Prabu, S. et al., "NLO-active Y-shaped ferrocene conjugated imidazole chromophores as precursors for SHG polymeric films," *Dalton Trans.*, vol. 49, no. 6, pp. 1854–1863, 2020.

[85] Milne, B. F. and P. Norman, "Resonant-convergent PCM response theory for the calculation of second harmonic generation in makaluvamines A-V: Pyrroloiminoquinone marine natural products from poriferans of genus Zyzzya," *J. Phys. Chem. A*, vol. 119, no. 21, pp. 5368–5376, 2015.

[86] Muhammad, S. "Second-order nonlinear optical properties of dithienophenazine and TTF derivatives: A butterfly effect of dimalononitrile substitutions," *J. Mol. Graph. Model.*, vol. 59, pp. 14–20, 2015.

Chapter 2

Modeling Efficient Second Harmonic Generation in Microcrystalline KDP Fibers – Part I

Matthew S. Neel*
Formerly with Oregon State University, Oregon, USA

Abstract

The phenomenon of second harmonic generation (SHG), or frequency doubling, in a relatively new (monoclinic) polymorphic structure of the well-known potassium dihydrogen phosphate (KDP) crystal was examined. While the literature on both the SHG and KDP is extensive, this new structure presents something of a mystery in that it can produce highly efficient harmonic signals and yet, paradoxically, it appears to have a relatively small nonlinear coefficient. Herein, we explore the foundations of SHG and the general properties of KDP in Part I, and then implement this knowledge in a numerical model presented in Part II.

1. Introduction

Optics, as a subject area within physics as a whole, is the study of a portion of the electromagnetic spectrum that we call *visible* light (for example, the colors of rainbow) and what we might call *near-visible* light or light that we cannot see with our eyes, but that is close enough to the visible range that its interactions are of interest to us, e.g., infrared and ultraviolet. We might

* Corresponding Author's Email: msn2700@gmail.com.

In: Second Harmonic Generation
Editors: Abdel-Baset M. A. Ibrahim and Pankaj Kumar Choudhury
ISBN: 978-1-68507-888-1

include other *invisible* light such as x-rays, gamma rays, radio waves, etc., but these are often treated separately. Optics is especially interested in the interaction of light with matter, such as reflection from mirrors, focusing using lenses, refraction of light moving from air to water, and the formation of images from sources of light. But, it also seeks to describe the nature of light itself, such as photons streaming through space. We understand light to have both wave and particle descriptions, and in many cases, one description is more successful (though not necessarily more intuitive) than the other in modeling the behavior of light.

We use the term *linear optics* to describe interactions in which an increase in light intensity of the source has a directly proportional effect on the intensity of the reflected or refracted light. Linear optics also typically deals with relatively low-intensity light sources (such as common light bulbs, some LEDs, filtered sunlight, etc.) and their interaction with objects having little or no symmetry at the atomic level (e.g., water, glass, air, and other amorphous solids and gases, etc.). In linear optics, light causes no real macroscopic change to the object that it shines upon; the object is mostly a passive participant. We might say that linear optics describes about 99% of our everyday experience with light, and the principles behind those experiences have been familiar to humanity for centuries.

The last 1% involves unusual interactions that have only been truly explored in the last few decades. This is the area known as *nonlinear optics* in which an increase in light intensity of the source can result in an exponential change in the intensity of the output light. Nonlinear optics often employs high-intensity sources (typically lasers or laser diodes) and studies their interaction with objects having a high degree of symmetry and purity at the atomic level (such as crystals, thin films, and optical fibers). In nonlinear optics, the incoming light is so intense that it causes real macroscopic changes to the target object. The object is thereby energized and becomes an active participant in the interaction. Such intense light can simply cause the bound electrons in a crystal to wiggle back and forth a little more violently than normal or, in extreme cases, can pour such a torrent of energy into the crystal that the heat causes it to expand and crack.

Nonlinear optics is also the realm of the microscopic in which, for most interactions, events at the atomic level must be modeled in detail. Classical electromagnetic theory, as well as quantum mechanics, are brought to bear and the consistency of their conclusions examined. At the atomic level, molecules can behave like radio antennas that absorb and reradiate different frequencies in preferred directions. This is especially true inside a crystal where the regular lattice of molecules will allow the movement of bound electrons more easily in one direction than another. When this ensemble of radio antennas reradiate energy together, it can give rise to bizarre effects in which incident light waves can be added (sum frequency generation), subtracted (difference frequency generation), and even multiplied (harmonic generation). We can make a clearer distinction between linear and nonlinear optics based on how the electric field from an incoming light wave *polarizes* the atoms and molecules within a material. In this usage, polarization refers to the degree to which an external electric field causes positive and negative charges within a material to separate. In many materials, the arrangement of charges is more or less random. This means that a particular charge is equally free to move in any direction and, to an electromagnetic wave, every direction "looks" the same; we call such materials *isotropic*.

In certain materials like specific classes of crystals, the arrangement of charges is not at all random, and a particular charge may be able to move a large distance in one direction, but a very short distance in another direction. In this case, an electromagnetic wave will travel differently based on its direction, and we call such materials *anisotropic*. Linear optics, therefore, deals with linear polarization, meaning that an increase in the external field will lead to a proportional increase in charge separation. Nonlinear optics deals with nonlinear polarization, meaning that an increase in the external field may lead to an exponential increase in charge separation. Also important is the property of *centrosymmetry*, or the ability to rotate a crystal structure into itself around an inversion center. We will see that *noncentrosymmetric* crystals exhibit the nonlinear properties necessary for second harmonic generation (SHG). It is this harmonic generation that we are interested in discussing in this chapter.

Aggregating microscopic phenomena into macroscopic effects requires just the right interplay between the incident laser light and the actively participating object. Many restrictions are placed on both of these partners for the output harmonic signal to be significant. The incident light, called the fundamental, must have the correct wavelength, adequate coherence, sufficient power density, appropriate polarization, and even the right input angle to the crystal. The crystal must, among the other properties, have the right molecular structure, be transparent at both the fundamental and harmonic wavelengths, have sufficient birefringence, and it must be strong enough to avoid laser damage from overheating. If we can ensure all of these elements are present and aligned experimentally, only then do we have a chance of inputting, for example, invisible infrared light into the crystal and seeing the dazzling green light that results.

As we might imagine, creating the right input light is less of a problem than finding the right crystal for the harmonic signal we would like to see. Laser technology has been developed to the point that a large variety of sources are available with a wide range of wavelengths, power, and operating modes. And while the discovery and development of new nonlinear crystals over the last few decades have been prolific, it can still be a challenge to find a crystal with the requisite atomic transition structure to produce a desired, arbitrary wavelength of light for a given input wavelength. Often this consists of selecting a common crystal host, such as KTP or BBO, and doping it with atoms that have the necessary energy levels. High purity crystals are also typically grown epitaxially, or one atomic layer at a time, which is a slow and expensive process even for volumes less than one cubic centimeter. Here we confine our study to the crystal KDP which is, next to quartz, one of the best-known nonlinear crystals. While its structure and capability for SHG as a bulk crystal is well-studied, our interest lies in its behavior in a new and unusual geometry.

In the pages that follow, we examine the well-known phenomenon of SHG in a new and unique setting. Recent efforts [1,2,3] by a crystal engineering research group have led to the non-equilibrium, ambient growth of a remarkable crystalline structure previously thought only to exist in extreme conditions [4]. When exposed to intense laser light, it displays the characteristics of a bulk, nonlinear crystal as well as a hollow-core, fiber optic waveguide. We will examine the foundations of the phenomenon from Maxwell's equations, describe KDP and the novel crystal structure in detail, and finally develop a 2D numerical model using geometric optics that allows us to simulate guided and radiation modes. The details of these modes will illustrate the potential that this unique structure has to offer to nonlinear optical applications.

2. Nonlinear Optics

2.1. Electron in an Asymmetric Potential Well

In the subsequent derivations, we generally follow Zernike and Midwinter's excellent treatment [5] and begin with the smallest part of this elaborate system: the Lorentz model of a single electron bound to an atom. In that model, a charged particle is bound by one field, but free to move in response to an applied field when the latter has sufficient strength to *almost* overcome the former (by completely overcoming, it would mean *ionization*). This atom lies in an anisotropic crystal that is immersed in an external electric field and, as such, the electron is free to move further in some directions than others. We start with the typical equation of motion for a damped harmonic oscillator in symmetric potential well with no applied force [6]

$$\frac{d^2r}{dt^2} + 2\gamma\frac{dr}{dt} + \omega_0^2 r = 0 \tag{2.1}$$

We include damping due to the molecular forces on the electron that seek to keep it in a particular spot. For small perturbations, the electron will oscillate back and forth, and radiate energy with an equation of motion given by eq. (2.1). Now, we turn on an external electric field which applies a force to the electron. With the field turned on, the shape of the well changes. We now include an anharmonic term to reflect the fact that the electron is in an asymmetric potential well. Putting all of these together, we have the modified expression

$$\frac{d^2r}{dt^2} + 2\gamma\frac{dr}{dt} + \omega_0^2 r - \xi r^2 = -\frac{e}{m}E \tag{2.2}$$

where r is the displacement of the electron from its equilibrium position, e is the electron charge, m is the electron mass, ω_0 is its natural oscillation frequency, γ is the damping constant, E is the externally applied electric field and ξ is a constant that defines the strength of the anharmonic term. Since we assume the anharmonic term will be small compared with the harmonic term, we can use a power series approach in which we break up the displacement into pieces and express that piecemeal distance in relation to the field strength that causes it, i.e.,

$$r = r_1 + r_2 + r_3 \cdots \tag{2.3}$$

$$r_i = c_i E^i \tag{2.4}$$

Since the physics we are concerned with relates to the square of the electric field, we keep only the first two terms and neglect the other (infinite) terms. We also allow the external field to have many components, i.e., it may not be monochromatic. The amplitude, phase, and frequency of each component are independent. So, the overall field can be written as a finite sum in the form

$$E = \sum_n E(\omega_n) e^{-i\omega_n t} \tag{2.5}$$

Notice that in eq. (2.5), there is no dependence on any spatial coordinates; we will address that later. For now, we are just highlighting the multi-component nature of the field. Inserting Eq. (2.3) into Eq. (2.2), we have

$$\frac{d^2}{dt^2}(r_1 + r_2) + 2\gamma \frac{d}{dt}(r_1 + r_2) + \omega_0^2 (r_1 + r_2) = \xi (r_1 + r_2)^2 - \frac{e}{m} E \tag{2.6}$$

We can separate this equation into two equations by collecting terms with similar power of *E*. Doing so leads to the pair of equations

$$\frac{d^2 r_1}{dt^2} + 2\gamma \frac{dr_1}{dt} + \omega_0^2 r_1 = -\frac{e}{m} E \tag{2.7}$$

$$\frac{d^2 r_2}{dt^2} + 2\gamma \frac{dr_2}{dt} + \omega_0^2 r_2 = \xi r_1^2 \tag{2.8}$$

where we have chosen to neglect the $2r_1 r_2$ and r_2^2 terms in the expansion of $(r_1 + r_2)^2$ since they are proportional to E^3 and E^4, respectively, and will thus be even smaller than thus be even smaller than the magnitude of phenomena in which we are interested. Thus eq. (2.7) represents the linear response of the system and eq. (2.8) represents the nonlinear response. Now we can use eq. (2.5) in eq. (2.4), and then take the result and use it in eq. (2.7) to express it in terms of sums over the field components. After taking the time derivatives, we have

$$\sum_n \{\omega_0^2 - 2i\omega_n \gamma - \omega_n^2\} c_1 E(\omega_n) e^{-i\omega_n t} = -\frac{e}{m} \sum_n E(\omega_n) e^{-i\omega_n t} \tag{2.9}$$

Since both sides of the equation are under the same sum, we can divide both sides by the $\{\cdots\}$ portion. We find the following expression for the linear displacement of electron

$$r_{linear} = r_1 = -\frac{e}{m}\sum_n \frac{E(\omega_n)e^{-i\omega_n t}}{\omega_0^2 - 2i\omega_n\gamma - \omega_n^2} \tag{2.10}$$

To find the nonlinear displacement r_2, we will need an expression for the square of electric field given in eq. (2.5), which we represent as

$$\left(\sum_n E(\omega_n)e^{-i\omega_n t}\right)^2 = \sum_n\sum_m E(\omega_n)E(\omega_m)e^{-i(\omega_n+\omega_m)t} \tag{2.11}$$

This is meant to show that, when we multiply two sums, we multiply every term in one sum by every term in the other, and maintain the distinction between members of the two sets by employing different indices. Returning to Eq. (2.8), let us take each term in sequence, for clarity, and express all of them in terms of the electric field

$$\frac{d^2r_2}{dt^2} = 2c_2\left[E\cdot\frac{d^2E}{dt^2} + \left(\frac{dE}{dt}\right)^2\right] \tag{2.12}$$

$$2\gamma\frac{dr_2}{dt} = 4\gamma c_2 E\frac{dE}{dt} \tag{2.13}$$

$$\omega_0^2 r_2 = \omega_0^2 c_2 E^2 \tag{2.14}$$

$$\xi r_1^2 = \xi(c_1E)^2 \tag{2.15}$$

Each term involves the square of the field, which is as expected. Using these values in eq. (2.8) and performing the time derivatives, leads to the double summation on both sides. This, in turn, allows us to gather terms as we did previously with $\{\cdots\}$ and divide both sides by it, which leads us to

$$r_{nonlinear} = r_2 = -\frac{e^2\xi}{m^2}\sum_n\sum_m \frac{E(\omega_n)E(\omega_m)e^{-i(\omega_n+\omega_m)t}}{F(\omega_0,\omega_n,\omega_m,\gamma)} \tag{2.16}$$

where the function in the denominator includes all the relevant frequencies (and their combinations) as well as the damping coefficient in one expression

$$F(\omega_0,\omega_n,\omega_m,\gamma) = (\omega_0^2 - 2i\omega_n\gamma - \omega_n^2)(\omega_0^2 - 2i\omega_m\gamma - \omega_m^2) \times [\omega_0^2 - 2i(\omega_n + \omega_m) - (\omega_n + \omega_m)^2] \tag{2.17}$$

2.2. External Electric Field and Induced Polarization

We are interested in how the external electric field affects the electron in question, and so, we need an expression for the polarization that the external field causes. Since the polarization density is a function of charge separation r, and we are assuming that all of our charges will respond to the external field in unison, we can similarly represent the polarization magnitude as a sum of many components, i.e.,

$$P = P_1 + P_2 + P_3 + \cdots = -Ner \tag{2.18}$$

where N is the electron density and e represents the charge of an electron. Since we know that r is also an expansion, we can state eq. (2.18) more concisely with a summation

$$P = \sum_{l=1}^{\infty} P_l = -Ne\sum_{i=1}^{\infty} r_i \tag{2.19}$$

Substituting in our expressions for the various r components and keeping the first two terms, as we did with the displacement expansion previously, we can represent the first two components of polarization as in the following

$$P_{lin} = P_1 = \sum_n \chi^{(1)}(\omega_n)E(\omega_n)e^{-i\omega_n t} \tag{2.20}$$

$$P_{nonlin} = P_2 = \sum_n \sum_m \chi^{(2)}(\omega_n, \omega_m) E(\omega_n) E(\omega_m) e^{-i(\omega_n + \omega_m)t} \tag{2.21}$$

It is worth pointing out that the linear polarization depends on a single electromagnetic wave while the nonlinear polarization represents, in general, the interaction of several electromagnetic waves. In this way, we can express the essential idea that the polarization of crystal responds in different ways to different waves. We also simplified the expressions by defining the electric susceptibility χ for the first and second-order polarizations, used in Eqs. (2.20) and (2.21), as

$$\chi^{(1)}(\omega_n) = \frac{Ne^2}{m} \frac{1}{\omega_0^2 - 2i\omega_n\gamma - \omega_n^2} \tag{2.22}$$

$$\chi^{(2)}(\omega_n, \omega_m) = -\frac{m\xi}{N^2e^3} [\chi^{(1)}(\omega_n)][\chi^{(1)}(\omega_m)][\chi^{(1)}(\omega_n + \omega_m)] \tag{2.23}$$

Up until now, our notation has been *scalar* and reflected the 1D case. But, now we treat the polarization and electric field as 3D vectors moving through a macroscopic crystal. To do so, we need to update our notation in eq. (2.21) and specifically call out individual components of the nonlinear polarization

$$P_i(\omega_{n+m}) = \sum_{jk} \sum_{mn} \chi_{ijk} E_j(\omega_n) E_k(\omega_m) e^{-i(\omega_n + \omega_m)t} \tag{2.24}$$

where *i*, *j*, *k* can each take on values for *x*, *y*, *z* directions, respectively, and, compared with the notation used previously, we call out the specific dependence on each frequency as well their combination, and are changing symbols so that $\chi^{(2)}(\omega_n, \omega_m) \rightarrow \chi_{ijk}(\omega_{n+m}, \omega_n, \omega_m)$. The susceptibility is now a third rank tensor and a function of individual frequencies involved and (more to the point) the interference between them. Its role is to store all the 3D information about how susceptible the crystal is (i.e., how to react) to an externally applied field to produce a polarization wave. For the interaction of three optical waves, there can be as many as 324 complex numbers in the tensor. For SHG with contracted notation and Kleinman symmetry, the susceptibility tensor has a maximum of 10 independent elements [7] depending on the type of crystal being described. As we will see, some crystal classes will have even less.

2.3. Maxwell's Equations and the Wave Equation

With the nonlinear polarization wave in eq. (2.24) as a source term, we can invoke Maxwell's equations to describe what is happening within the crystal. As the charges are accelerated and jiggle back and forth, we know they emit electromagnetic radiation in all directions. We have no free currents in the crystal, no net charges in the crystal, and our crystal is nonmagnetic. Thus, Maxwell's equations [8] reduce to these (3D vectors are indicated in bold):

$$\nabla \times \boldsymbol{H} = \frac{1}{c}\frac{\partial \boldsymbol{D}}{\partial t} \tag{2.25}$$

$$\nabla \times \boldsymbol{E} = -\frac{1}{c}\frac{\partial}{\partial t}(\mu \boldsymbol{H}) \tag{2.26}$$

$$\mathbf{D} = \varepsilon \boldsymbol{E} + 4\pi \boldsymbol{P}_{nl} \tag{2.27}$$

We note that $\boldsymbol{P}_{nl}$ here refers to the nonlinear polarization *only* since the linear polarization is accounted for in the first term on the right-hand side of eq. (2.27). Taking the curl of both sides of Eq. (2.26) and then substituting Eqs. (2.25) and (2.27) into the result gives us the familiar wave equation

$$\nabla^2 \boldsymbol{E} = \frac{\partial^2}{\partial t^2}\left(\frac{\varepsilon}{c^2}\boldsymbol{E}\right) - \frac{4\pi}{c^2}\frac{\partial^2 \boldsymbol{P}_{nl}}{\partial t^2} \tag{2.28}$$

This is a general rule that we will apply to our specific situation. Here we assume that various waves travel in only the $\hat{z}$-direction so that all other spatial derivatives become zero for simplicity. We also assume that there are three monochromatic electromagnetic waves present within the crystal, and we define those individually as

$$E_1(z,t) = E_1(z)e^{-i(\omega_1 t - k_1 z)}$$
$$E_2(z,t) = E_2(z)e^{-i(\omega_2 t - k_2 z)}$$

$$E_3(z,t) = E_3(z)e^{-i(\omega_3 t - k_3 z)} \tag{2.29}$$

We imagine that two of these waves are input into the crystal, and the third wave is generated by the crystal in response to them. We also define each polarization component individually in a similar way and note that, according to eq. (2.21), each component arises from the interaction of two of the three electromagnetic waves

$$\begin{aligned} P_{1z}(z,t) &= 4dE_2^*(z)E_3(z)e^{-i[(\omega_3-\omega_2)t-(k_3-k_2)z]} \\ P_{2z}(z,t) &= 4dE_3(z)E_1^*(z)e^{-i[(\omega_3-\omega_1)t-(k_3-k_1)z]} \\ P_{3z}(z,t) &= 4dE_1(z)E_2(z)e^{-i[(\omega_1+\omega_2)t-(k_1+k_2)z]} \end{aligned} \tag{2.30}$$

where, for example, $k_1^2 = \varepsilon\omega_1^2/c^2$ is the square of the wave vector, and we have arranged our terms to reflect a sum frequency setup in which we have $\omega_1 + \omega_2 = \omega_3$. In other words, the crystal will react in such a way that the frequency of the response will be the sum of the frequencies of the input waves. Note that we are using the convention for the nonlinear coefficient when the Kleinman symmetry is valid, or $d = \chi_{ijk}/2$.

Now that we have each of the pieces defined, we can insert them into our wave equation. It is straightforward to perform the second-order time derivatives from the right-hand side of eq. (2.28) on the components shown in Eqs. (2.29) and (2.30). For the Laplacian (second-order gradient) on the left-hand side of eq. (2.28), we see the full expression would be, using the first field and first polarization components as an example, such as

$$\frac{\partial^2 E_1(z,t)}{\partial z^2} = \left[-k_1^2 + 2ik_1\frac{\partial}{\partial z} + \frac{\partial^2}{\partial z^2}\right]E_1(z)e^{-i(\omega_1 t - k_1 z)} \tag{2.31}$$

where the term in square brackets operates on the field $E_1(z)$. We assume, as is typical, that the third term inside the brackets is much smaller than the first and second terms (once they operate on the field) and can be ignored. Ignoring the $\partial^2/\partial z^2$ term is known as the *slowly varying amplitude* approximation [9] in which we assume that the amplitude is not subject to rapid changes like shock waves or square/sawtooth pulses. Given that we are interested in continuous wave (CW) operation of the pump light rather than pulsing, this is a fair assumption to make. We repeat this substitution process for the second and third field/polarization components. This leads to three coupled amplitude equations that express the interaction of the three waves as

$$\frac{dE_1(z)}{dz} = -i\frac{8\pi\omega_1^2}{k_1c^2}dE_2^*(z)E_3(z)e^{-i(k_3-k_2-k_1)z}$$
$$\frac{dE_2(z)}{dz} = -i\frac{8\pi\omega_2^2}{k_2c^2}dE_3(z)E_1^*(z)e^{-i(k_3-k_2-k_1)z}$$
$$\frac{dE_3(z)}{dz} = -i\frac{8\pi\omega_3^2}{k_3c^2}dE_1(z)E_2(z)e^{-i(-k_3+k_2+k_1)z} \tag{2.32}$$

We have so far assumed, and will continue to assume, that there are no losses inside our various mediums. These coupled equations fully describe the interdependence of each wave on the other two, and they are still generalized such that they can be applied to any of several situations.

3. Second Harmonic Generation

3.1. SHG Coupled Amplitude Equations

We now have the essential mathematical setup for sum-frequency generation and we now turn to the classic special case of it: SHG or frequency doubling. Since SHG requires an intense, coherent light source, its discovery followed closely upon the invention of laser [10,11]. In this phenomenon, both the input electromagnetic waves have the same frequency and they generate a wave with twice the frequency, or $\omega_1 + \omega_1 = 2\omega_1 = \omega_2$. We often summarize this by saying that the medium absorbs two photons of the same energy and emits a single photon having twice the energy of either of the inputs. We can imagine two people holding each end of a single rope and whipping it with the same motion to produce two incoming wave pulses. During the "collision" the combined wave packet has twice the frequency and half the wavelength. The energy of the waves overall, assuming no losses, is of course conserved.

In this situation, the coupled amplitude equations in eq. (2.32) reduce to a simpler set since we do not need to distinguish between three separate waves, just the input wave (twice) and the resulting harmonic:

$$\frac{dE_1(z)}{dz} = -i\frac{8\pi\omega_1^2}{k_1c^2}dE_1^*(z)E_2(z)e^{-i\Delta kz}$$
$$\frac{dE_2(z)}{dz} = -i\frac{16\pi\omega_1^2}{k_2c^2}dE_1^2(z)e^{i\Delta kz} \quad (3.1)$$

We define here the phase mismatch $\Delta k = 2k_1 - k_2$ between the generated wave and the incoming waves. Notice that this is basically the difference between the relevant wave vectors, and it will be multiplied by the path length into the crystal to calculate how far away from each other the relevant waves will have drifted by the time they arrive at the given crystal depth. Thus, the phase mismatch is an important measure of spatial coherence, which is necessary for the waves to exchange energy. If the value of Δk is large, the waves are moving in fundamentally different directions and will not cohere. We can imagine that the fundamental and harmonic waves are two trains that run along tracks. As long as the tracks are very close together and parallel, cargo (energy) can be tossed back and forth across the gap by the train crew as they move. But, if the tracks begin to diverge (nonzero phase mismatch), the exchange is no longer possible since the trains have drifted away from each other. Only if the phase mismatch is nonzero can the fundamental "train" provide energy to power and sustain the harmonic "train."

3.2. Approximations and Implications

Many sources in the literature use the *small-signal approximation* (also called the un-depleted pump approximation), or the assumption that the amplitude of the generated wave is small enough (the energy conversion is low enough) that the amplitudes of the two input waves are essentially constant as they move through the length of the crystal. We specifically do *not* make that assumption here, and thus, allow for a potentially large second harmonic signal; we will need it later on (one of the necessary assumptions of our numerical model will be that a large harmonic wave may always be present once we are deep enough into the crystal). One way to proceed is to write out the real and imaginary parts of the equations in Eq. (3.1). Note that the amplitude of the wave can be written to reflect the fact that the wave has a phase that is dependent on distance or $\varphi(z)$, i.e.:

$$E_1(z) = \frac{1}{2}\mathcal{E}_1(z)e^{i\varphi_1(z)} \tag{3.2}$$

Doing so and then separating the real and imaginary parts via Euler's formula leads us to the following set of equations:

$$\frac{d\mathcal{E}_1}{dz} = -\frac{4\pi\omega^2 d}{c^2 k_1}\mathcal{E}_1\mathcal{E}_2 \sin\theta \tag{3.3}$$

$$\frac{d\mathcal{E}_2}{dz} = \frac{8\pi\omega^2 d}{c^2 k_1}\mathcal{E}_1^2 \sin\theta \tag{3.4}$$

$$\frac{d\theta}{dz} = \Delta k - \frac{8\pi\omega^2 d}{c^2}\left[\frac{\mathcal{E}_2}{k_1} - \frac{\mathcal{E}_1^2}{\mathcal{E}_2 k_2}\right]\cos\theta \tag{3.5}$$

$$\theta = \Delta kz + 2\varphi_1(z) - \varphi_2(z) \tag{3.6}$$

where Eqs. (3.3) and (3.4) are the changes in amplitude with depth into the crystal, eq. (3.5) is the change in phase mismatch with depth into the crystal and eq. (3.6) is the general expression of the overall phase mismatch itself. Note that eq. (3.6) has two parts: one for the wave vector mismatch and the other which we will call the "traveling" phase mismatch. From the Manley-Rowe relations [7] (one form of which we give here) we have:

$$\frac{n_1 c}{\omega_1}\frac{d}{dz}(E_1 E_1^*) = \frac{n_2 c}{\omega_2}\frac{d}{dz}(E_2 E_2^*) = -\frac{n_3 c}{\omega_3}\frac{d}{dz}(E_3 E_3^*) \tag{3.7}$$

which describes the conservation of energy among three interacting electromagnetic waves as they travel within the crystal. Applying this to the case of SHG, we have for the power flow as:

$$W = W_1 + W_2 \tag{3.8}$$

$$W = \frac{k_1 c^2}{8\pi\omega_1}\mathcal{E}_1^2 + \frac{k_2 c^2}{8\pi(2\omega_1)}\mathcal{E}_2^2 \tag{3.9}$$

where W is the total initial power directed into the crystal, which might be a combination of both the fundamental and harmonic lights. And substituting Eqs. (3.3) and (3.4) into Eq. (3.5) yields:

$$\frac{d\theta}{dz} = \Delta k + \left(\frac{\cos\theta}{\sin\theta}\right)\frac{d}{dz}[ln(\mathcal{E}_1^2\mathcal{E}_2)] \tag{3.10}$$

At this point, it is typical to make assumptions and solve Eqs. (3.3) and (3.4) using any of the various techniques. Normally it is assumed that $\Delta k = 0$ (i.e., perfect phase-matching) and that there is no harmonic signal at the start of the spatial interaction, or $\mathcal{E}_2(0) = 0$. To keep our derivation as general as possible (for reasons that will become clear), we will *not* be making either of those assumptions. The most general case in the literature (which avoided those assumptions) was solved early on [12] and we follow it here. We define a characteristic length:

$$L = \left[\frac{4\pi\omega_1^2 d}{k_1 c^2}\mathcal{E}_1(0)\right]^{-1} \tag{3.11}$$

This is the length over which the fundamental wave shifts about 70% of its energy to the harmonic wave when perfectly phase-matched. In practice we want the characteristic length to be much greater (e.g., 1000×) than the wavelength of the fundamental wave for efficient harmonic conversion [13]. It is worth pointing out that the larger the initial amplitude of the fundamental wave, the shorter the characteristic length. Since shorter and smaller crystals are less expensive and easier to grow than larger crystals, using a shorter crystal places a requirement on the input beam to have a higher power density, and therefore, a shorter characteristic length, in order to ensure significant transfer of the fundamental energy to the harmonic wave.

3.3. A Fully Generalized Solution

Armstrong, Bloembergen, Ducuing, and Pershan [12] derived the most general solution for harmonic power as a function of crystal depth, and here we follow their work. Since this solution is generally not applicable in typical SHG experiments in the laboratory, it is absent from all nonlinear optics textbooks and can be found only in the original paper. We need to make some changes to simplify our expressions and update our notation before seeking our main solution. We switch to normalized amplitudes by using the following:

$$u(\zeta) = \mathcal{E}_1(z)\sqrt{k_1c^2/8\pi\omega_1 W} \tag{3.12}$$

$$v(\zeta) = \mathcal{E}_2(z)\sqrt{k_2c^2/16\pi\omega_1 W} \tag{3.13}$$

Expressed in this way, conservation of power flow eq. (3.9) has the simple and intuitive form of summing to unity:

$$u^2 + v^2 = 1 \tag{3.14}$$

We have also adjusted our distance variable from *z* using:

$$\zeta = Cz \tag{3.15}$$

Where the adjustment constant *C* contains all the appropriate constants as well as the input power *W*, and is given by:

$$C = (4\pi\omega_1^2 d/k_1c^2)\sqrt{16\pi\omega_1 W/k_2c^2} \tag{3.16}$$

Using these new definitions, we can rewrite our original differential Eqs. (3.3), (3.4), and (3.5) as:

$$\frac{du}{d\zeta} = -uv\sin\theta \tag{3.17}$$

$$\frac{dv}{d\zeta} = u^2\sin\theta \tag{3.18}$$

$$\frac{d\theta}{d\zeta} = \frac{\Delta k}{C} + \left(\frac{\cos\theta}{\sin\theta}\right)\frac{d}{d\zeta}[ln(u^2v)] \tag{3.19}$$

We begin with the last of the three equations, or eq. (3.19), and, at first, we assume perfect phase-matching so that $\Delta k = 0$. This allows us to setup the equation for integration:

$$\int \left(\frac{\sin\theta}{\cos\theta}\right) d\theta = \int d\ [ln(u^2 v)] \tag{3.20}$$

Doing so leads to a constant of integration that we define as:

$$\Gamma = u^2 v \cdot \cos\theta_0 \tag{3.21}$$

This is similar to finding the homogeneous solution of the original differential equation. Going back to eq. (3.19) and now assuming a nonzero phase mismatch, or $\Delta k \neq 0$, we perform a similar integration with this result for the inhomogeneous case:

$$u^2\, v \cdot \cos\theta + \frac{1}{2} v^2(\zeta) \frac{\Delta k}{C} = u^2 v \cdot \cos\theta_0 + \frac{1}{2} v^2(0) \frac{\Delta k}{C} \tag{3.22}$$

Gathering terms with common factors, we can rewrite Eq. (3.22) as:

$$v(1 - v^2)\cos\theta = \Gamma - \frac{1}{2}\frac{\Delta k}{C}[v^2 - v^2(0)] \tag{3.23}$$

Since we are already taking into account the initial traveling phase mismatch, we do not need to do it twice and can set $\theta = 0$ in eq. (3.23). We are seeking an expression for the harmonic power as a function of depth into the crystal, but it turns out to be easier to invert that and instead find a crystal depth for a given harmonic power. Since the harmonic power is expected to oscillate, this leads to many different solutions for a given power. To that end, we can square both sides of eq. (3.23), subtract one side from the other, and then take the square root in root-mean-square fashion to ensure the correct sign and magnitude. This converts it into the important function:

$$G(v^2) = \left[v^2(1 - v^2)^2 - \left\{\Gamma - \frac{1}{2}\frac{\Delta k}{C}[v^2 - v^2(0)]\right\}^2\right]^{1/2} \tag{3.24}$$

The argument v^2 is the normalized harmonic power, but this function also uses $\Gamma, \Delta k$ and the initial harmonic power $v^2(0)$ as parameters. This allows us to set up the differential equation for the change in harmonic power as a function of path length:

$$(d/d\zeta)(v^2) = \pm 2G(v^2) \tag{3.25}$$

The right-hand side will be positive or negative depending on whether the sign of the initial phase mismatch is positive or negative, respectively. The general solution for the path length into the crystal can be given in terms of elliptic integrals:

$$\zeta = \pm\frac{1}{2}\int_{v^2(0)}^{v^2(\zeta)} \frac{d(v^2)}{G(v^2)} \tag{3.26}$$

3.4. Harmonic Power and Energy Exchange

We leave the discussion of elliptic integrals to the original authors [12], who examined those in detail. The key is that the equation $G(v^2) = 0$ has three roots, which are often found graphically, and v^2 is constrained to move between those three values, which gives rise to the oscillatory behavior of energy that we expect is being exchanged between the fundamental and the harmonic waves. By setting the parameters in Eq. (3.24), we set the stage for a unique solution for the integration, and thereby for the harmonic signal. Returning to our original notation, there is a condition for imperfect phase matching, i.e.:

$$\Delta kL \cdot \mathcal{E}_2(0) \gg \mathcal{E}_1(0) \tag{3.27}$$

which compares the initial amplitude of the fundamental to the initial amplitude of the harmonic magnified by the wave-vector mismatch. If this condition is met, we can then express the solution to the original coupled amplitude Eqs. (3.3)–(3.4) as:

$$\theta = \Delta kz + \theta_0 + \pi/2 \tag{3.28}$$

$$\mathcal{E}_2(z) = \mathcal{E}_2(0) + \frac{2}{\Delta kL}\mathcal{E}_1(0) \times [\sin(\theta_0 + \Delta kz) - \sin\theta_0] \tag{3.29}$$

And simply taking the square of eq. (3.29), we have:

$$\mathcal{E}_2^2(z) = \mathcal{E}_2^2(0) + \frac{4}{\Delta kL}\mathcal{E}_1(0)\mathcal{E}_2(0) \times [\sin(\theta_0 + \Delta kz) - \sin\theta_0] + \frac{4}{(\Delta kL)^2}\mathcal{E}_1^2(0) \times [\sin(\theta_0 + \Delta kz) - \sin\theta_0]^2 \tag{3.30}$$

This is a key result and will determine much of the behavior of our numerical model. The structure of eq. (3.30) implies that the harmonic signal has three important parts: the initial harmonic power, an exchange term that involves the product of both amplitudes modulated by the sine of mismatch, and the initial fundamental power modulated by the square of the sine of mismatch. We can insert eq. (3.30) into Eq. (3.9) to get the corresponding equation for $\mathcal{E}_1(z)$. We notice from eq. (3.30) that the amplitude of harmonic oscillates as a function of depth in the crystal as the energy is exchanged back and forth between the fundamental and harmonic [11,7]. Under *very* precise conditions and ultra-pure crystals, harmonic conversion can theoretically reach >99%, but high conversion often requires complex experimental setups, such as several crystals chained together [14] or carefully constructed input beam profiles [15], in order to manage the inherent dispersion, losses and beam divergence.

4. The KDP Crystal

4.1. Sellmeier Equations, Effective Index and Phase-Matching

Recall that all the optical phenomena of which a nonlinear crystal is capable stem from the natural asymmetry in its molecular structure that gives bound electrons more freedom to move in one direction than in another, i.e., an asymmetric potential well. To observe nonlinear optical effects, it goes without saying that we will need a nonlinear crystal. Described as "the most extensively studied nonlinear optical crystal in history" [1], potassium dihydrogen phosphate (KH_2PO_4) or KDP is relatively easy to grow and has been the industry standard reference crystal for nonlinear optical measurements for decades [16]. As a negative uniaxial crystal, it has a single optical axis such that, if a ray travels along that axis, it experiences no birefringence (refractive effects). Since nonlinear crystals behave very differently than the lenses and mirrors described in most of the optics-based courses (since these elements are centrosymmetric), and since our numerical model will pertain specifically to KDP, it is worth our time to review (briefly) some of the relevant equations and their applications. Derivations can be found in the listed references. Here we merely highlight the key equations that will be used in our model.

In optics, the index of refraction is the ratio of the speed of light in a vacuum to the speed of light in that material or $n = c/v$. In general, light with a shorter wavelength will travel more slowly than light with a longer wavelength. For every wavelength, a generic nonlinear crystal has (at least) two indices of refraction depending on how the light ray is plane-polarized relative to the optical axis: an ordinary index of refraction and an extraordinary index of refraction. In a negative crystal, the former is larger than the latter for a given wavelength. The Sellmeier equations for KDP, which relate the index of refraction to the wavelength of light in question, are given by [17]:

$$n_o^2(\lambda) = 2.259276 + \frac{0.01008956}{\lambda^2 - 0.012942625} + \frac{13.00522\lambda^2}{\lambda^2 - 400} \quad (4.1)$$

$$n_e^2(\lambda) = 2.132668 + \frac{0.008637494}{\lambda^2 - 0.012281043} + \frac{3.2279924\lambda^2}{\lambda^2 - 400} \quad (4.2)$$

where λ is the wavelength in the micron scale.

The speed of an electromagnetic wave (its phase velocity) traveling through the crystal depends not only on its direction but also on its polarization and wavelength. In general, an anisotropic crystal will only allow plane wave propagation in one of the two orthogonal directions. If a ray has ordinary polarization, its phase velocity will be the same; no matter what direction it travels within the crystal. If a ray has extraordinary polarization, its phase velocity will depend on its angle with the optical axis. In the latter case, the index of refraction that the wave "sees" will have a value between $n_o(\lambda)$ and $n_e(\lambda)$ called the *effective index*, given by [5]:

$$n_{eff}(\theta,\lambda) = \frac{n_e(\lambda)\, n_o(\lambda)}{\sqrt{(n_o(\lambda)\,\sin\theta)^2 + (n_e(\lambda)\,\cos\theta)^2}} \quad (4.3)$$

where θ is the angle between the light ray and optical axis, and should not be confused with the general phase mismatch defined earlier in eq. (3.6). To calculate what is called the *phase-matching angle*, or the angle between the fundamental wave vector and the optical axis, for perfect phase matching in SHG and for a negative uniaxial crystal-like KDP [7], we have:

$$\theta_m = \sin^{-1}\left(\sqrt{\left[\frac{1}{n_o^2(\lambda_1)} - \frac{1}{n_o^2(\lambda_2)}\right] \Big/ \left[\frac{1}{n_e^2(\lambda_2)} - \frac{1}{n_o^2(\lambda_2)}\right]}\right) \tag{4.4}$$

For example, $\theta_m = 41.207^\circ$ for a typical KDP crystal when $\lambda_1 = 1064$ nm and $\lambda_2 = 532$ nm. There is another graphical way to find the phase-matching angle that also involves looking at the various indices of refraction, and that is frequently shown (but rarely described in detail) in the literature [18]. First, we choose a point anywhere inside our crystal and orient ourselves by finding the optical axis. We must find a direction to travel in 3D space for which $n_o(\lambda_1) = n_{eff}(\theta_m, \lambda_2)$; this is the essence of angular phase-matching. We imagine a circle of radius $n_o(\lambda_1)$ centered at our point (with the optical axis along its diameter) and an ellipse with a shape determined by $n_{eff}(\theta_m, \lambda_2)$ centered at the same point (with the optical axis along its minor axis).

These two shapes will intersect at four points, and if we measure the angle θ_m between the optical axis and a ray that connects our central point to any of the four intersection points, we will find the phase-matching angle. Since this is a 2D construction, we rotate the entire picture along the optical axis to sweep out the 3D construction. Our four possible rays thereby sweep out two cones whose vertices meet at the central point we choose; this makes intuitive sense since the intersection of a concentric sphere and ellipsoid will result in two plane-parallel circles, which form the edges of cones. See figure. 1 for details. It is important to point out that *anything* that changes the inter-molecular distances within the crystal will affect the indices of refraction, and thereby the ability to phase-match, including heating/cooling the crystal, applying a voltage across the crystal, or applying mechanical strain.

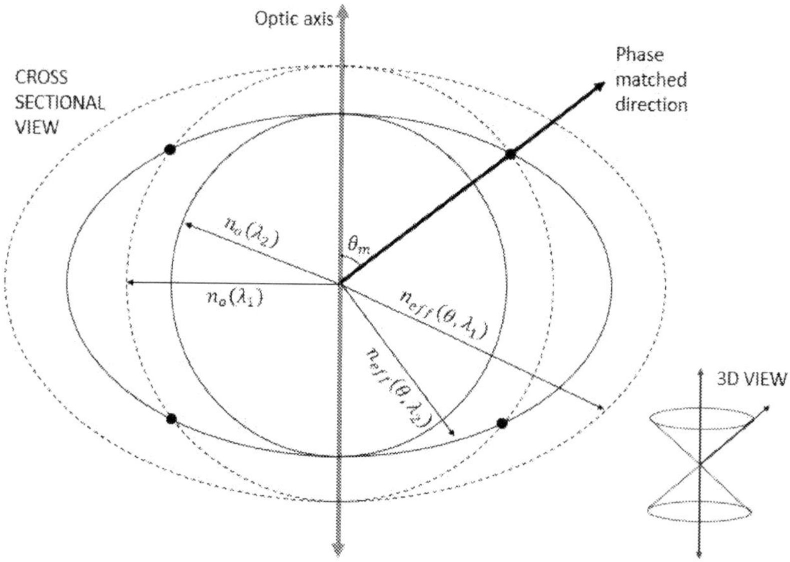

Figure 1. Graphical construction to find the phase matching angle; solid lines refer to the harmonic, dashed lines refer to the fundamental; the four intersection points are shown along with some additional lines for illustrative purposes.

4.2. Snell's Law, Fresnel Coefficients and KDP Structure

When light moves from vacuum into a linear medium like glass at an angle, we typically use Snell's law to determine the relevant angles. The familiar form of Snell's law relates the angle of incidence θ_i to the angle of transmission θ_t, and we note that the angle of incidence is equal to the angle of reflection θ_r, i.e.:

$$n_i(\lambda)\sin\theta_i = n_t(\lambda)\sin\theta_t$$

$$\theta_i = \theta_r \tag{4.5}$$

But, in the case of nonlinear crystals, the index of refraction in the next medium (the crystal) *depends on* the angle of transmission such that we have:

$$n_i(\lambda)\sin\theta_i = n_t(\theta_t,\lambda)\sin\theta_t \tag{4.6}$$

Setting up and solving this equation is difficult because of the many directions involved, and is usually performed graphically. The nonlinear version of Snell's law [5] to find the transmission angle for a light ray incident upon a negative uniaxial crystal like KDP, after a somewhat elaborate geometrical construction (different from the more typical Huygens' construction), is given by:

$$\cot\theta_t = \frac{(o^2 - e^2)\sin 2\alpha + 2\sqrt{e^2a^2/i^2 + o^2b^2/i^2 - o^2e^2}}{2(e^2a^2 + o^2b^2)} \tag{4.7}$$

where $e = 1/n_e(\lambda)o = 1/n_o(\lambda)$, , $i = \sin\theta_i$, $a = \sin\alpha$, $b = \cos\alpha$, α is the angle that the optical axis makes with the interface and θ_i is the angle of incidence. Once we know and can insert all the necessary values, we are able to recover the angle of refraction. For KDP, the linear and nonlinear versions of Snell's law produce results that differ by $\leq 1°$ for any orientation of the optical axis, but for other crystals, the difference can be more significant. Depending on the experimental setup being modeled and the precision required, accurate ray geometry may require using eq. (4.7).

As a nonlinear crystal, KDP is generally anisotropic and non-centrosymmetric and, as mentioned previously, the strength of nonlinearity that a light ray experiences while traveling through the crystal depends on its direction and polarization. The strength is represented by d, which we encountered in eq. (2.30), and the exact expression of d is specific for each crystal class, phase matching type, and symmetry. For the Type 1 phase matching in KDP (when Kleinman symmetry is valid), we have [19]:

$$d_{T1}(\theta,\varphi) = d_{36}\sin(\theta + \rho)\sin(2\varphi) \tag{4.8}$$

where d_{36} is the element from the 3rd row, 6th column of the nonlinear electric susceptibility tensor for KDP, θ is again the angle that the input ray makes with the optical axis of the crystal, ρ is the birefringent walk-off angle and φ is the angle between the plane containing the input ray and the optical axis and the a-axis of the crystal. The birefringent walk-off angle [20], which is related to double refraction, is the angle of separation between two beams that enter a crystal collinearly, and diverge as they travel through the crystal due to their different wavelengths and polarizations. The coefficient d_{36} sets the general strength of nonlinearity for the various processes; the larger the value, the greater the harmonic signal will be. We can express the relationships between the three spatial polarization components, the susceptibility tensor and the products of the electric field magnitudes in matrix form (specifically adapted for SHG) using the contracted index form [7] as:

$$\begin{bmatrix} P_x(2\omega) \\ P_y(2\omega) \\ P_z(2\omega) \end{bmatrix} = 2 \begin{bmatrix} d_{11} & d_{12} & d_{13} & d_{14} & d_{15} & d_{16} \\ d_{21} & d_{22} & d_{23} & d_{24} & d_{25} & d_{26} \\ d_{31} & d_{32} & d_{33} & d_{34} & d_{35} & d_{36} \end{bmatrix} \begin{bmatrix} E_x(\omega)^2 \\ E_x(\omega)^2 \\ E_x(\omega)^2 \\ 2E_y(\omega)E_z(\omega) \\ 2E_x(\omega)E_z(\omega) \\ 2E_x(\omega)E_y(\omega) \end{bmatrix} \tag{4.9}$$

Notice that all of the elements in the electric field matrix are products of two amplitudes, though *which* two amplitudes are involved varies by row. When Kleinman symmetry is valid, only three values of the 18 elements in the tensor are non-zero and $d_{14} = d_{25} = d_{36}$. Typical values for KDP are $d_{36} = 0.44\text{pm/V}$ [17] and $\rho = 1.60°$ [19]. This is the expression of *d* that would be used in the coupled amplitude equations described previously. And, as we can see from those equations, this nonlinear coefficient partially determines the strength of the nonlinear response. As a note, Kleinman symmetry relates diagonal components of the 18 elements in the tensor above and states that they may be equal as long as the frequency of the input electromagnetic waves is far away from any natural resonances in the material that may lead to asymmetric behavior [21].

Using the correct phase-matching angle (referred to as the critical phase-matching) is a part of the process, but there is still a choice of polarization of the two fundamental waves that generate the harmonic wave. We assume the fundamental waves can be linearly polarized in one of the two perpendicular orientations: the ordinary (o) or the (e) extraordinary direction, or parallel and perpendicular to the optical axis, respectively. When using a negative uniaxial crystal, the harmonic wave will always be e-polarized. If both the fundamental waves are o-polarized, we call this the Type 1 phase-matching. If one of the fundamentals is o-polarized and the other is e-polarized, we call this the Type 2 phase-matching. Sometimes this is summarized in the literature with three symbols such as "*ooe*" (Type 1) or "*oee*" (Type 2) to stand for "fundamental, fundamental, harmonic" polarization [18].

Since polarization is important to SHG and the behavior of electromagnetic waves as they reflect off or refract into the crystal, we must keep careful track of the energy stored in the fundamental and harmonic waves for both the s-type and p-type polarizations. The polarization known as the s -type refers to an electric field that is polarized perpendicular to the plane of incidence; p-type refers to a field whose polarization is parallel to the plane of incidence. To ensure that the energy is conserved with both reflection and refraction (as well as in special situations like total internal reflection), we calculate the power reflection and transmission Fresnel coefficients for a vacuum/dielectric interface [22]. Let us assume the light ray is moving from vacuum into the KDP crystal. After we have determined the angles of incidence and refraction, we can use the following to find the reflected power:

$$R_s = \left|\frac{n_{vac}\cos\theta_i - n_{KDP}\cos\theta_t}{n_{vac}\cos\theta_i + n_{KDP}\cos\theta_t}\right|^2 \tag{4.10}$$

$$R_p = \left|\frac{n_{KDP}\cos\theta_i - n_{vac}\cos\theta_t}{n_{KDP}\cos\theta_i + n_{vac}\cos\theta_t}\right|^2 \tag{4.11}$$

Once we find the reflected power, we can easily find the transmitted power using the relations $T_s = 1 - R_s$ and $T_p = 1 - R_p$.

KDP is a member of the $\bar{4}2m$ crystal class, and generally forms tetragonal-shaped (scalenohedral) crystals. But, KDP is known for its polymorphisms, or the different crystal structures that it can form via nucleation once the aqueous solution is brought up to temperature and then allowed to cool [23]. With extreme conditions, such as high ambient pressure, KDP can be forced into structures that would not form (and could not be maintained at standard conditions) otherwise. One such structure, known as the Type IV polymorphism, was first obtained at the pressures on the order of 1–2 GPa (>9800+ atm) and temperatures greater than 400K [4]. This special shape is a monoclinic crystal that forms long fibers with a typical length to diameter ratio of 500:1. These microcrystalline fibers have a hexagonal cross-section and occur as both the hollow-core and solid-core. The underlying crystal cells self-assemble along their relatively short b-axis to form these hexagonal fibers. The diameters of these fibers range from less than one micron to tens of microns, most likely with various wall thicknesses for the hollow core variety.

One research group [1] was able to grow these identical structures at room temperature and pressure. The exact mechanism is, as of yet, unknown. Even stranger is the fact that the crystal is almost centrosymmetric and, as such, should be a linear crystal with no nonlinear properties. Despite that, it efficiently generates second harmonics. Examination via x-ray diffraction reveals a twin-crystal structure resulting in a very weak nonlinear coefficient. This occurs when what appears to be a single crystal lattice is, in fact, two lattices that are very nearly superimposed in space. When sufficiently intense laser light is input into one end of the fiber, the effect of this small nonlinearity is somehow many-times magnified, leading to relatively high harmonic conversion with an efficiency of the order of $\eta = P_{2\omega}/P_{\omega} \approx 10^{-4}$. It is this fundamental mystery that we wish to explore with a numerical reference model in Part II of this study.

Conclusion

We have reviewed the basic theory and mathematical formalism of SHG and the role that wave vector mismatch plays in the generation of optical harmonics. We focused on finding the most general solution to the coupled amplitude equations without making assumptions a priori that would eliminate important physical phenomena. Given that we will be exploring a new structural variant of KDP, we wish to keep all phenomena available from a modeling perspective. In the next chapter, we will use the formulae we derived here to build a numerical model that will help to explain the new and exciting behavior of this KDP variant.

References

[1] Ren, Y., X. Zhao, E. W. Hagley, and L. Deng, "Ambient-condition growth of high-pressure phase centrosymmetric crystalline KDP microstructures for optical second harmonic generation," *Sci. Adv.*, Vol. 2, No. 8, 2016.

[2] Ren Y. and S. J. Zhanga, "Optical quality tetragonal phase single-crystal fiber of potassium di-hydrogen phosphate with efficient second-harmonic generation," *Cryst Eng Comm*, vol. 19, pp767-771, 2017.

[3] Ren, Y., C. Ma, C. C. S. Chan, C. Q. Zhang, and K. S. Wong, "Second harmonic generation from UV to visible in KDP single-crystalline fibers," *Conference on Lasers and Electro-Optics OSA Tech*. Dig. (online), p. JTu2A.133, 2018.

[4] Caia W. and A. Katrusiak, "Structure of the high-pressure phase IV of KH_2PO_4 (KDP)," *Dalton Trans.*, vol. 42, pp. 863-866, 2013.

[5] Zernike F. and J. E. Midwinter, "*Applied Nonlinear Optics,*" Dover Publications reprint (2006) of Wiley, New York, NY (1973)

[6] Marion J. B. and S. T. Thornton, "*Classical Dynamics of Particles and Systems,*" 4th Ed., Harcourt College Publishers, New York, USA (1995).
[7] Boyd, R. W. "*Nonlinear Optics,*" 2nd Ed., Academic, New York, NY (2003).
[8] Jackson, J. D. "*Classical Electrodynamics,*" 3rd Ed., Wiley, New York, NY (1999).
[9] Kleinman, D. A. "Theory of second harmonic generation of light," *Phys. Rev.*, vol. 128, No. 4, pp. 1761-1775, 1962.
[10] Franken, P. A., A. E. Hill, C. W. Peters, and G. Weinreich, "Generation of optical harmonics," *Phys. Rev. Lett.*, vol. 7, No. 4, pp. 118-120, 1961.
[11] Kuzyk, M. G. "*Nonlinear Optics: A Student's Perspective,*" CreateSpace Independent Publishing Platform, North Charleston, SC (2017).
[12] Armstrong, J. A., N. Bloembergen, J. Ducing, and P. S. Pershan, "Interactions between light waves in a nonlinear dielectric," *Phys. Rev.*, vol. 127, No. 6, pp. 1918-1939, 1962.
[13] Shen, Y. R. "*The Principles of Nonlinear Optics,*" Wiley, New York, NY (1984).
[14] Kumar, S. C., G. K. Samanta, Kavita Devi, and M. Ebrahim-Zadeh, "High-efficiency, multicrystal, single-pass, continuous-wave second harmonic generation," *Opt. Express,* vol. 19, pp. 11152–11169, 2011.
[15] J. Arlt, K. Dholakia, L. Allen, and M. J. Padgett, "Efficiency of second-harmonic generation with Bessel beams," *Phys. Rev. A*, vol. 60, No. 3, pp. 2438-2441, 1999.
[16] N. Bloembergen, "*Nonlinear Optics: A Lecture Note and Reprint Volume,*" Frontiers in Physics, W. A. Benjamin Inc. Publishers, New York, NY (1965).
[17] Castech Inc., KDP Product Sheet, http://www.castech.com (2017).
[18] Midwinter J. E. and J. Warner, "The effects of phase matching method and of uniaxial crystal symmetry on the polar distribution of second-order non-linear optical polarization," *Br. J. Appl. Phys.*, vol. 16, no. 8, pp.1135-1142, 1965.
[19] Eckardt, R. C., H. Masuda, Y. X. Fan, and R. L. Byer, "Absolute and relative nonlinear optical coefficients of KDP, KD*P, BaB/sub 2/O/sub 4/, LiIO/sub 3/, MgO:LiNbO/sub 3/, and KTP measured by phase-matched second-harmonic generation," *IEEE J. of Quantum Electron.*, vol. 26, no. 5, pp.922-933, 1990.
[20] Smith A. V. et al. "Increased acceptance bandwidths in optical frequency conversion by use of multiple walk-off-compensating nonlinear crystals," *J. Opt. Soc. Am. B*, vol. 15, no. 1, pp. 122-141, 1998.
[21] Jia, M., X. Cheng, M. Whangbo, M. Hong, and S. Deng, "Second harmonic generation responses in KH2PO4: importance of K and breaking down of Kleinman symmetry," *RSC Adv.*, vol. 10, no. 44, pp. 26479–26485, 2020.
[22] Born M. and E. Wolf, "*Principles of Optics,*" 7th Ed., Cambridge Univ. Press, Cambridge, UK (1999).
[23] Rashkovich, L. N. "*KDP-Family Single Crystals,*" The Adam Hilger Series on Optics and Optoelectronics, IOP Publishing Ltd., New York, NY (1991).

Additional Resources

[1] Kleinman D. A. and R. C. Miller, "Dependence of second-harmonic generation on the position of the Focus," *Phys. Rev.*, vol. 148, no. 1, pp. 302-312, 1966.

[2] Hobden, M. V. "Phase-matched second-harmonic generation in biaxial crystals," *J. Appl. Phys.*, vol. 38, no. 11, pp. 4365-4372, 1967.

[3] Boyd, G. D. "Parametric interaction of focused Gaussian light beams," *J. Appl. Phys.,* vol. 39, no. 8, pp. 3597-3639, 1968.

[4] Chen, C. K., A. R. B. de Castro, and Y. R. Shen, "Surface-enhanced second-harmonic generation," *Phys. Rev. Lett.,* vol. 46, no. 2, pp. 145-156, 1981.

[5] Riesz R. P. and R. Simon, "Reflection of a Gaussian beam from a dielectric slab," *J. Opt. Soc. Am. A*, vol. 2, no. 11, pp. 1809–1817, 1985.

Chapter 3

Modeling Efficient Second Harmonic Generation in Microcrystalline KDP Fibers – Part II

Matthew S. Neel*
Formerly with Oregon State University, OR, USA

Abstract

Herein, the familiar phenomenon of second harmonic generation (SHG), or frequency doubling, in a relatively new (monoclinic) polymorphic structure of the well-known potassium dihydrogen phosphate (KDP) crystal was examined. A key factor in this may be the geometry of the structure itself, which self-assembles as a microscopic linear rod or tube with a hexagonal cross-section that has many of the properties of an optical fiber. In Part I, the SHG theory and KDP crystal properties were reviewed. In Part II, the potential roles that this geometry plays, was described considering the development and results of a numerical reference model that simulates SHG in an identical geometry but uses the more common (tetragonal) crystal lattice structure of KDP.

1. Introduction

In Part I we began with a general discussion of nonlinear optics and examined the details of creating a simple model for the interaction between an electron in an asymmetric well and intense laser light. Using this model as a spatially

* Corresponding Author's Email: msn2700@gmail.com.

In: Second Harmonic Generation
Editors: Abdel-Baset M. A. Ibrahim and Pankaj Kumar Choudhury
ISBN: 978-1-68507-888-1

repeatable unit cell, we expanded our discussion to a macroscopic crystal and how its atomic structure behaves like a phased antenna array that preferentially captures and converts specific wavelengths of light incident from specific directions. We also reviewed the specific optical formulae relating to a KDP crystal having the most common lattice structure. Second-harmonic generation, especially with a typical KDP crystal, is well understood.

What is not well understood, at least at first, is how newly synthesized crystal structures will behave in similar circumstances. Since atomic structure determines everything in regard to how a crystal will respond to intense electromagnetic radiation, a small change in even a common crystal-like KDP can lead to unexpected behavior. The optical equations we examined for KDP are *specific to a certain crystal structure* and, in general, cannot be applied to different atomic structures even for the same crystal material. From a scientific investigation standpoint, changing the atomic structure means a change to many variables simultaneously and deprives us of the ability to change and test a single variable at a time.

However, we have sufficient details on a new polymorphic structure of KDP to gain insight into how it produces efficient SHG light. A crystal's ability to produce harmonic light depends, at a macroscopic level, on the shape of the electron's asymmetric well and on how the bulk crystal was grown/cut. The former aspect is still being investigated and is beyond the scope of this discussion. The latter, though, is within reach of simulation since its shape is distinct and fairly easily modeled. As a result, we will take a kind of middle ground and assume the typical atomic structure of KDP but imagine that it is cut into the same shape as the naturally grown, polymorphic variety. In this way, we can examine what the effect of this geometry is on the production of harmonic light. And, as we will discover, the effect of geometry alone can be significant.

2. Numerical Model

2.1. Hexagonal Cross-Section of the KDP Fiber

We learned in Part I that research has demonstrated efficient SHG with a monoclinic crystal cell that self-assembles into a hexagonal macroscopic structure; we will focus on the hollow-core variety. Because the actual mechanism for harmonic generation in this new microcrystalline structure is still being investigated, we wish to build a reference model that describes how a typical KDP crystal would behave *if it was cut* to resemble this new hexagonal geometry. To better understand the physical mechanisms that may be leading to efficient creation of harmonics in this new structure, we built a 2D numerical model that assumes a typical KDP crystal (tetragonal cell), but with the same cross-sectional (hexagonal) shape and dimensions as the newly discovered structures. We assume that (i) the optical axis (the c-axis) as well as the a-axis of the crystal lies in this cross-sectional plane, (ii) the optical axis is at an arbitrary angle θ_{optic} to the laboratory x-axis, and (iii) the b-axis of the crystal points into the page along the length of the fiber. The nominal diameter of the fiber is 15 μm with a wall thickness of 3 μm which, unless specified otherwise, is what we will assume henceforward for modeling purposes (see Figures 1 and 2 for details).

We will model the classic SHG setup with a fundamental wavelength of $\lambda_1 = 1064$ nm and a harmonic wavelength of $\lambda_2 = 532$ nm. This translates to the wall thickness being $t \approx 3\lambda_1$ and $t \approx 6\lambda_2$, and a trip around the hexagonal, central perimeter would be $36\lambda_1$ and $72\lambda_2$. This presents a challenge in terms of the choice of approach for our model. The ray description of light generally does not apply when structures being considered are on the order of a wavelength [1]. Sub-wavelength structures, such as thin films, exhibit constructive and destructive interference, and are typically best represented by a wave description to capture the standing modes [2]. At the same time, for structures just a few wavelengths thick, wave and ray descriptions can have excellent agreement [3–5]. For that reason, we will use the ray approach. It is important to note that, while confined inside the crystal, the shortest ray path is 3 μm (i.e., the thickness of the hexagonal wall) and the longest is 12 μm.

In terms of basic assumptions, we assume that the crystal is placed in vacuum. For the benefit of simplicity at the possible expense of accuracy, we assume that all waves are infinite, monochromatic plane waves and the KDP crystal is lossless/has no resonances at the relevant frequencies. Obviously, we cannot have infinite plane waves existing in such a finite space in our crystal structure, but to maintain generality, we choose not to assume an intensity profile for the input laser beam. Our input beam is assumed to be CW rather than pulsed since we seek to model the steady-state operation. We also assume that the wave normal direction is approximately the same as the ray direction. Strictly speaking, this is not the case, but the structures are small enough that the difference in direction is slightly over such a short distance, and deviations average out over many reflections inside the structure. We also assume that our crystal can respond effectively instantly to an applied field such that we will not need to include the concept of retarded time in our numerical expressions. Physically, this just means that the response time is negligible compared to the transit time of a light ray across the crystal structure.

The "mechanics" of the geometric optics program are as follows. The details of the hexagonal structure are stored in three arrays: one for the x-coordinates of the vertices, one for the y-coordinates of the vertices and one for the face angles of the sides. The lengths of the hexagonal sides for both the inner and the outer surfaces (as well as other details) are stored in global constants that are referenced by various functions throughout the code, meaning that, with some small changes, the code could be adapted for hollow hexagons of other sizes, solid-core hexagons or even other cross-sectional shapes like triangular structures. The user specifies the coordinates and launch angle of the single, initial ray, which carries all of the input power. The initial ray can be launched from anywhere in the Cartesian plane: from within the hollow core of the hexagon, from within the crystal itself or from an arbitrary point far outside the structure. The path of that initial ray is then projected, and the code determines if it will impact a crystal/vacuum interface. If not, the ray ends just beyond an allowed range; if so, the impact point becomes the starting point of new possible rays, and the process iterates.

2.2. Ray Geometry and Processing Algorithms

The determination of whether the projected path of the initial ray will intersect any of the sides of the hexagonal structure is itself an interesting bit of code. It relies on a function that answers a very simple mathematical question: Given two-line segments with any length and orientation in the Cartesian plane, what are the coordinates of the intersection point (if it exists)? There are various solutions available online [6] and in computer science textbooks. Our algorithm implements it by calculating two values:

$$S_1 = \frac{(y_3 - y_4)(x_1 - x_3) + (x_4 - x_3)(y_1 - y_3)}{(x_4 - x_3)(y_1 - y_2) + (x_1 - x_2)(y_4 - y_3)} \tag{2.1}$$

$$S_2 = \frac{(y_1 - y_2)(x_1 - x_3) + (x_2 - x_1)(y_1 - y_3)}{(x_4 - x_3)(y_1 - y_2) + (x_1 - x_2)(y_4 - y_3)} \tag{2.2}$$

where the segment A has the end-points (x_1, y_1) and (x_2, y_2) and the segment B has the end-points (x_3, y_3) and (x_4, y_4). If we have $1 \geq S_1 \geq 0$ and $1 \geq S_2 \geq 0$, we can find the intersection point using S_1 in these expressions

$$x_{inter} = x_1 + S_1 \cdot (x_2 - x_1) \quad (2.3)$$

$$y_{inter} = y_1 + S_1 \cdot (y_2 - y_1) \quad (2.4)$$

The code employs these equations and looks for intersection points between the projected ray path, and all of the line segments making up the hexagonal structure. It may happen that the projected path will intersect multiple hexagonal line segments and, if this is the case, the program selects the closest intersection point as the impact location. If no intersection is found, the ray in question is allowed to travel a small distance for visibility in the accompanying plot, and is then ended.

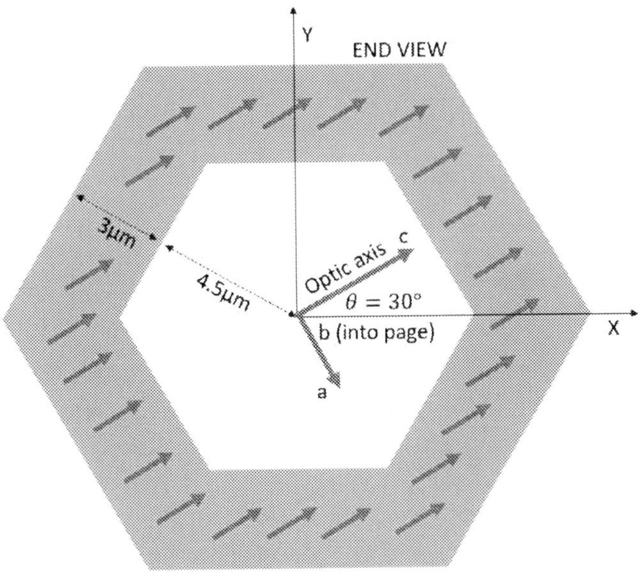

Figure 1. End-on view of the KDP microcrystalline fiber; the optical axis angle, measured from the x-axis is arbitrarily set to $\theta = 30°$ for illustration purposes (and represented within the crystal as darker arrows). The x- and y- axes shown are laboratory axes.

The growth of the harmonic wave is generally insignificant over distances on the order of a wavelength [7], so we expect a gradual build-up of harmonic energy as the fundamental wave bounces around in the hexagonal crystal. While the ray description usually neglects interference since it does not include phase properties, the model does track the accumulation of phase mismatch for the purpose of SHG. We include bulk crystal SHG in the model, and ignore surface generated SHG since the latter is generally many orders of magnitude smaller than the former. Our model has characteristics similar to both a hexagonal waveguide [8,9] and a hollow-core optical fiber [10,11]. As with any waveguide, our crystal undoubtedly has cutoff wavenumbers and allowed frequencies. As with all fiber optics, our structure demonstrates both the radiation and guided modes.

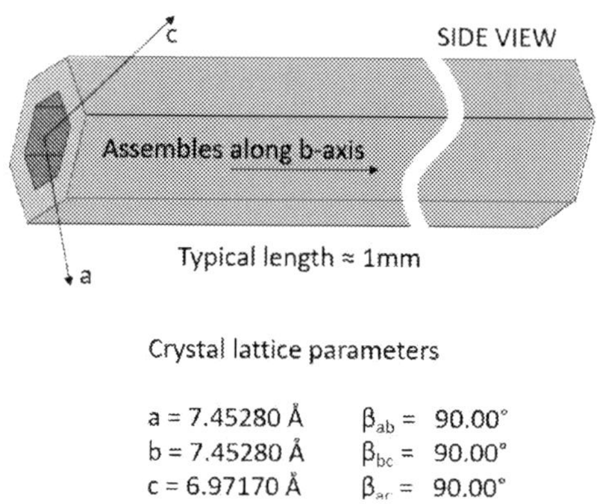

Figure 2. Side view of the KDP microcrystalline fiber; lengths and angles are given for the basic crystal lattice unit.

At the heart of the model is a ray creation and processing algorithm. The user initializes the properties of the first ray, and the ray propagates from its starting location to the point of next impact, e.g., contact with a vacuum-crystal interface. The initial ray can have both the fundamental and harmonic energies in arbitrary proportions. If the ray moves through bulk crystal, and if the conditions are favorable, the energy is exchanged between the fundamental and harmonic as SHG. If it moves through vacuum, it is unaffected. Depending on the impact conditions for that ray (the angle of incidence, the initial and final media, etc.), four rays might be created depending on the relevant Fresnel coefficients and surface SHG: (i) reflection of the fundamental, (ii) reflection of the harmonic, (iii) refraction of the fundamental, and (iv) refraction of the harmonic. These rays are added to a queue; the initial ray is then logged and removed from the queue. This ends the first iteration, and the program loops around to perform the same steps on the next ray in the queue. See Figure 3 for further details. The outputs of the process are a sort of miniature database of all rays created (their properties are shown as both the pre- and post- transit), and also, a plot showing up to 256 of these rays and their paths as they enter/exit the crystal.

In the program, we use a ray object model that can actually store properties of both the fundamental and the harmonic simultaneously. As a result, the reflected rays, which have the same reflection angle, are represented as a single ray object (rather than two) in the code to increase the speed of the algorithm. Thus, in each iteration of the loop, we are processing and removing one ray and adding *up to* three more. This process of "remove 1, create 3" leads to an ever-growing, potentially infinite queue of rays that must be processed to convey a complete picture of the movement and conversion of energy in the crystal. Therefore, there are two iterative limits placed on the algorithm. There is a queue limit, which halts the creation of rays as soon as the queue reaches the maximum number allowed (set by the user) and continues processing the rays on record until the queue is empty. There is also an iteration limit which halts the program entirely once a specified number of total rays have been processed. These can be used in combination to achieve the desired granularity of ray modeling. Separate from these two limits is an energy threshold that can be set be the user. For example, the user can set this threshold to ignore (prevent the creation of) any rays carrying less than a specified amount of energy.

Since our ray object model can accommodate four types of energy (*s*- and *p*- polarizations for both the fundamental and harmonic waves), we are able to capture much of the essential physics. From the Fresnel coefficients, we allow for Brewster's angle effects and total internal reflection, which occur for the incident critical angles $\theta_1 \geq 42.02°$ for the fundamental and $\theta_2 \geq 41.39°$ for the harmonic. Both the linear and nonlinear versions of Snell's law are included and applied as needed. Given the 2D nature of the model, we allow the Type 1 SHG (fundamental waves are ordinary or *s*-polarized; harmonic wave is extraordinary or *p*-polarized) for rays moving through the bulk crystal. We might say that bulk crystal SHG uses a $S_{\lambda_1} \rightarrow P_{\lambda_2}$ conversion mechanism, and this is the pathway that we will work. We also include the expected ≤180° phase change for reflection from an interface, where applicable.

2.3. Code Structure and Source

The code, which is fully commented with options and references, is written in VBA in a single MS Excel file of 1.1 MB. The run-time on an average laptop is only a few seconds to generate a ray database with 5,000 rays and the accompanying plot. The model is freely available on GitHub at https://github.com/msn2700/KDP-Hexagon. In terms of the organization of the code, there are three standard modules (Model, Plot and Test) and one class module (Ray). *Model* contains the main simulation algorithm and all relevant functions; *Plot* contains the code to generate the accompanying plot after the algorithm creates the data; *Test* is a simpler testing routine that can be used for calibration, quick checking and reference crystal experiments if desired. Within Model, there is a subroutine called Simulation that, when run, will create the small database of rays that result from the starting conditions and global properties. It outputs this database to one of the tabs in the spreadsheet.

The *Ray* class module stores the properties and methods of a ray object as we have defined it, including the four types of energy, *x*/*y* coordinates and propagation direction from the starting point, a unique ID and the ID of its parent ray if it was created as a result of reflection or refraction. The ray object also has three methods: it can clone itself; it can report out all of its properties, and it can update its properties post-transit through the crystal to reflect the occurrence of bulk SHG. Cloning of the rays involves a user-defined type (UDT) that acts as a template for all the properties of a ray. This template becomes one of the ray properties and provides it within a means to assign values to the other properties.

The Plot module handles the creation of the plot after Simulation has been run in the Model module. The GeneratePlot subroutine in the Plot module scans through the ray database that was created and collects the *x*/*y* data points of the rays to plot them as line segments on a 2D scatterplot with the hexagonal structure shown in the background. As an option, each segment can be assigned a color that is tied to a property of the ray. For example, ray segments can be colorized red with RGB values proportional to the total energy within the ray: darker red for higher energy and fading to lighter red for lower energy. In this way, the user can visualize how the energy disperses within the crystal with each reflection/refraction as more and more rays are created.

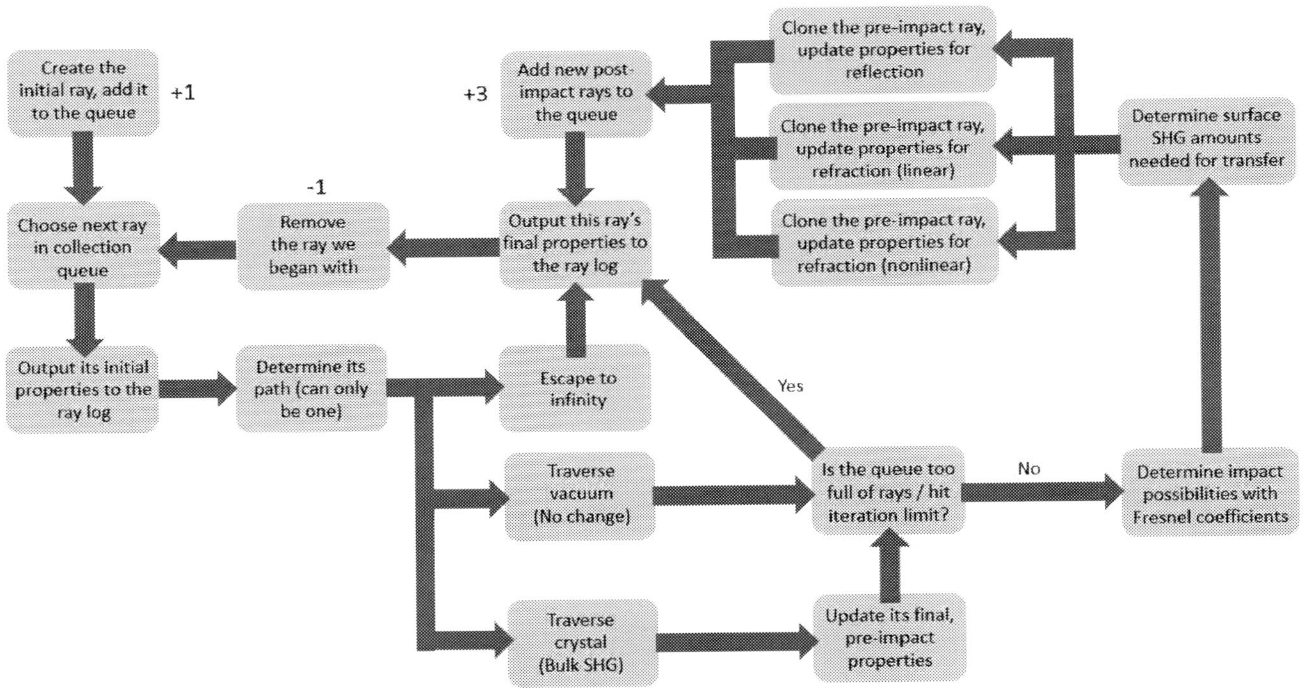

Figure 3. The algorithm for the 2D model; the program ends when the queue is depleted of rays (all rays have been processed and logged). Generating the plot using the ray database is a separate program module.

As with any geometrical optics program, there are many relative angles used in the cadre of functions; so, it is worth describing the general angular coordinate system. Some angles are measured relative to the optical axis (such as the phase-matching angle); others are measured with respect to a normal (such as the angle of incidence or angle of refraction). The absolute angular reference is the standard one for Cartesian coordinates, i.e., for a ray with its tail on the origin, we measure the angle defining its direction as counterclockwise from the $+x$-axis. Without exception, the code uses degrees rather than radians and converts to radians where necessary. When a function in the code describes the input angle as *standard form*, this is what is meant by that. When a ray is incoming, and will reflect off of one of the sides of the hexagon, the program determines which of the two normal directions to that surface is more appropriate to use in finding the angle of incidence.

3. Model Validation

3.1. Hallmarks of the SHG Signal

Before applying our 2D numerical model to the novel geometry in question, we validated the model by simulating a typical SHG setup using a common KDP crystal. Given the wealth of information in the literature relating to this crystal and its use in SHG phenomena, it was relatively simple to find various plots and figures of merit that we could replicate to ensure that the underlying functions were working properly. It must be pointed out that demonstrating SHG using KDP does not imply a single, standard experimental setup; there are many different setups depending, for example, on the choice of Type I or Type II polarization, CW or pulsed laser pump, temperature tuning of the crystal, etc. As a result, we show merely that our numerical model can demonstrate the same general behaviors that one might see in any of these setups without perfectly replicating the results of any of them. Given that our model makes simplifying assumptions, such as a lossless crystal, it would be impossible to perfectly replicate any real-world experiment with our simulations.

For these test simulations, we assume normal incidence of a high-intensity, plane wave laser beam on a KDP crystal that can be cut at any desired angle. We are free to set the crystal to an arbitrary thickness. Our laser beam can also carry both the fundamental and harmonic waves in any proportion collinearly such that both can be incident on the crystal with a specified initial phase mismatch. We are also free to set the spatial step size of simulation to much less/greater than a fundamental wavelength. As we might imagine, this gives us a great deal of freedom to simulate even non-physical phenomena but, to remain realistic, we perform several critical tests on the model to ensure accurate representation. For most of the figures that follow, we use normalized power to give a sense of scale.

3.1.1. Energy Exchange Between Fundamental and Harmonic

Recall that we deviated from most textbook derivations of SHG by allowing for three possibilities: a depleted pump, a non-zero phase mismatch and initial non-zero harmonic signal. This has important consequences in that we will not be able to compare our results with typical textbook relations [12] that predict, for example, the following relationships for the fundamental and harmonic amplitudes at a depth z in the crystal

$$\mathcal{E}_2(z) = \mathcal{E}_1(0)\tanh(z/L) \tag{3.1}$$

$$\mathcal{E}_1(z) = \mathcal{E}_1(0)\,\mathrm{sech}(z/L) \tag{3.2}$$

where L is defined, as is typical, to represent the length over which the fundamental wave shifts about 70% of its energy to the harmonic wave when perfectly phase-matched [12],

$$L = \left[\frac{4\pi\omega_1^2 d}{k_1 c^2}\mathcal{E}_1(0)\right]^{-1} \tag{3.3}$$

The authors have made three assumptions: an un-depleted pump (small harmonic signal), $\mathcal{E}_2(0) = 0$ and $\Delta k = 0$. These expressions predict saturation at depths of a few multiples of L. Because we allow for a non-zero phase mismatch, the power flow oscillates between the fundamental and harmonic fields provided that the crystal is long enough to allow this to happen (usually many times the characteristic length L). As such, our main validation is to demonstrate this periodic energy exchange and ensure that the total energy is conserved. We see the basic energy exchange we were seeking in Figure 4. When looking at the shape of the oscillations, we see a portion with a larger radius of curvature (fundamental peaks, harmonic troughs) and a portion with a smaller, more pinched radius of curvature (fundamental troughs, harmonic peaks). This indicates the presence of even and odd harmonics characteristic of a nonlinear, non-centrosymmetric medium [13].

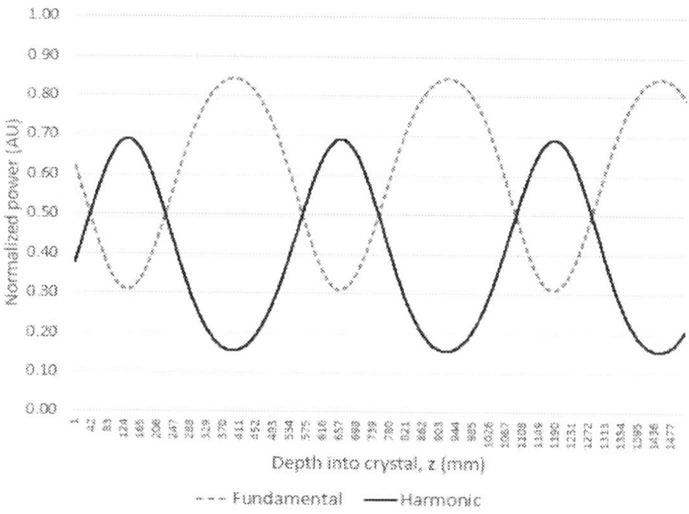

Figure 4 – Oscillation of energy between the fundamental and harmonic in an infinitely long crystal for various depths; $\theta = 41.207°$, step size $\Delta z = 0.5$ mm and initial, incident power proportion is $P_{2\omega}(0)/P_{\omega}(0) = 3/5$.

3.1.2. Sensitivity to Beam Angle

The sensitivity of the second harmonic signal on the direction of the fundamental ray through the crystal is well known [14]. Deviations of only $\Delta\theta = \pm 1°$ from the phase matching angle can lead to drastically reduced harmonic signals in many experimental setups, given the sensitivity of the phase mismatch on direction. The asymptote in Figure 5 corresponds to perfect phase-matching and, by virtue of dividing by Δk in the harmonic amplitude equation [12]:

$$\mathcal{E}_2(z) = \mathcal{E}_2(0) + \frac{2}{\Delta k L}\mathcal{E}_1(0) \times [\sin(\theta_0 + \Delta k z) - \sin\theta_0] \tag{3.4}$$

to an "infinite" harmonic signal (i.e., dividing by a value that approaches zero). The harmonic signal is less sensitive to changes $\Delta\varphi$, and peak conversion corresponds to $\varphi = 45°$ as expected from the $\sin(2\varphi)$ term in the expression for the nonlinear coefficient for Type I phase matching in KDP[7]:

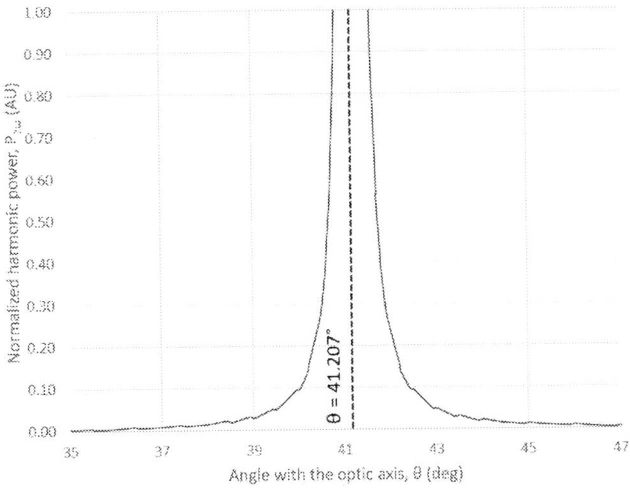

Figure 5 – Sensitivity of the harmonic signal on θ; the power decreases rapidly to zero for even small deviations from $\theta_m = 41.207^\circ$; note that the "jagged" appearance of the lines at the base of the peak is an artifact of the plotting software.

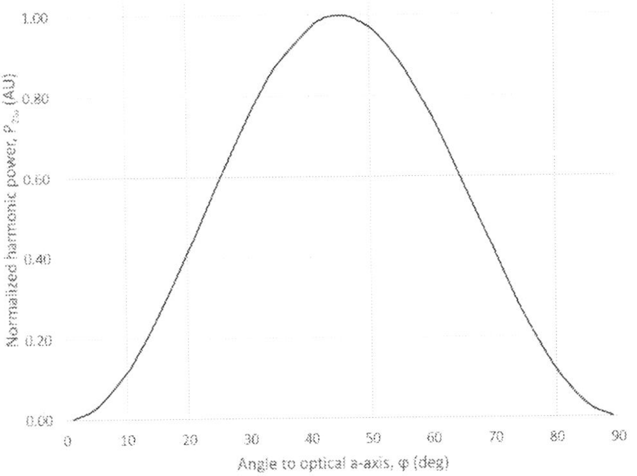

Figure 6 – Variation of the harmonic signal with changes in φ; there is less sensitivity to this angle compared with fig. 5.

$$d_{T1}(\theta, \varphi) = d_{36} \sin(\theta + \rho) \sin(2\varphi) \tag{3.5}$$

This is shown in Figure 6.

3.1.3. Demonstration of Maker Fringes

A common, early technique in the 1970s of measuring the nonlinear optical coefficient of materials was the production of Maker fringes using SHG [15,16]. It essentially measured the interference between the bound and free harmonic waves within the crystal as a function of path length through the crystal. Experimentally this is accomplished by rotating a rectangularly shaped crystal along the optical axis and watching the harmonic signal exiting the crystal varying with rotation angle. Since the input laser beam is perpendicular to the optical axis during the rotation, the same degree of angular phase-matching is maintained throughout. Since the path length through the crystal is a function of that angle, this is effectively the same as varying the thickness of the crystal, something that would otherwise be impossible to do without creating an entire array of crystals with different thicknesses. When the thickness is varied via the rotation, the phase contribution given by Δkz to the overall phase mismatch [12] between fundamental and harmonic waves,

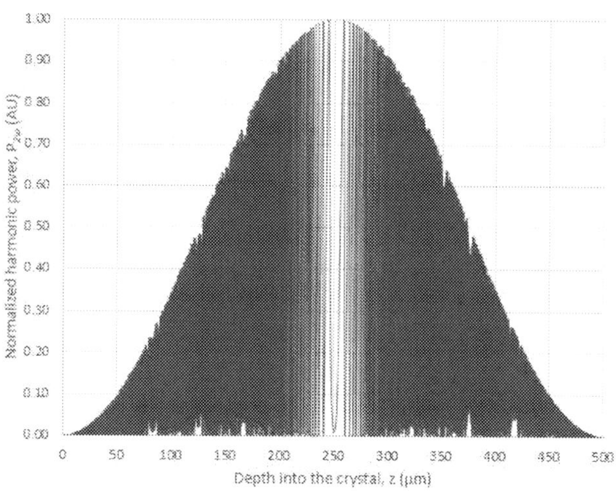

Figure 7 – Maker fringes for a hypothetical KDP crystal with $\theta = 39.75°$ and step size $\Delta z = 100$ mm; the independent variable measures depth into the crystal with normal incidence.

$$\theta = \Delta kz + \theta_0 + \pi/2, \tag{3.6}$$

is varied. Since our numerical model can have an arbitrary step size and propagation angle, it is relatively simple to simulate Maker fringes with various parameters. The results are shown in Figures 7 and 8.

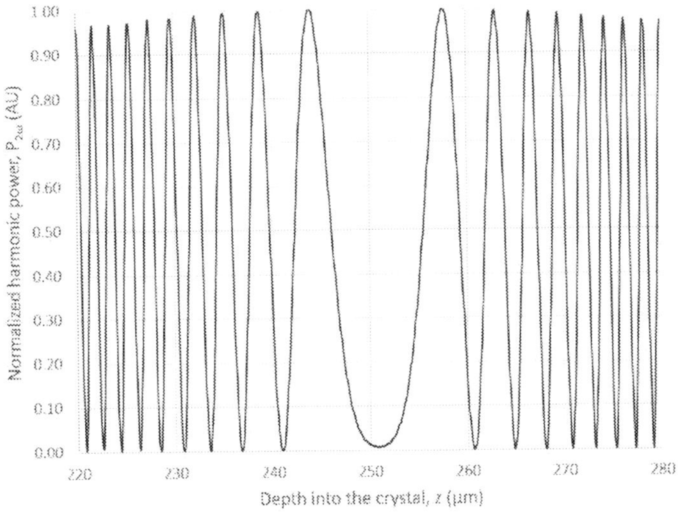

Figure 8. An enlarged view from Figure 7 showing the rapidly oscillating nature of harmonic with variation in depth as well as the central null at a path depth of 250 μm.

3.1.4. Harmonic Amplitude Variation with Phase Mismatch

As was shown previously, the amplitude of harmonic wave, given by eq. (3.4) above, has a dependence like $\sin(\Delta kz)/\Delta kL$. As such, the value of the phase mismatch will effectively determine the period as well as magnitude of the signal as it travels through the crystal. As $\theta \to \theta_m$, the phase mismatch approaches zero, or $\Delta k \to 0$. When this happens, the amplitude increases and the period decreases as the harmonic signal saturates and full conversion becomes possible. Figure 9 shows this effect for three example values of θ relatively far from the phase-matching angle. The three signals begin in-phase, but their different periods lead to different amplitude shapes, giving rise to the "bouncing" behavior commonly seen in the literature. When perfect phase-matching occurs, according to theory, the harmonic amplitude builds up to 100% conversion in the limit of an infinitely long crystal. In that case, the amplitude has the profile of a hyperbolic tangent that quickly builds, and then approaches maximum as an asymptote.

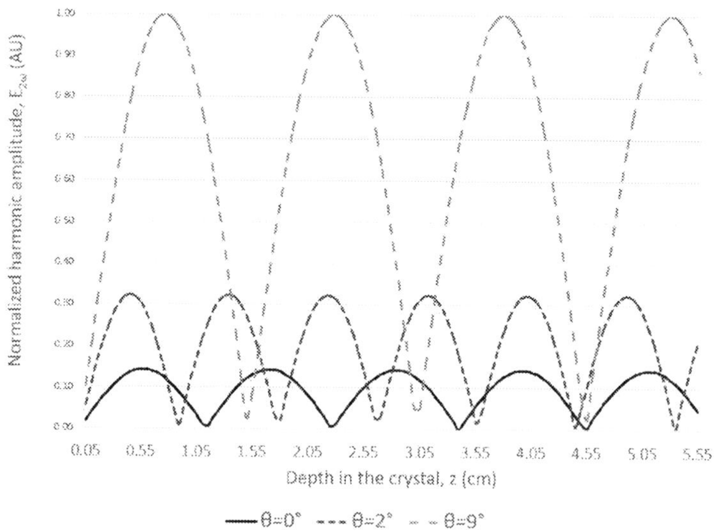

Figure 9. Apparent "bouncing" of the harmonic amplitude with depth into a hypothetical (long) KDP crystal; as the angle moves toward the phase-matching angle, the amplitude increases exponentially.

4. Simulation Results

4.1. Radiation and Guided Modes Inside the Crystal

The research group hypothesized that the underlying mechanism of their harmonic conversion was a kind of effective, Type 1 quasi-phase matching process in which the "angularly distributed pump … guarantees that there will always be a certain portion of the pump that satisfies the Type 1 phase matching" direction [17]. As we have seen, a positive wave vector mismatch, or $\Delta k > 0$, leads to a back-and-forth exchange of energy between the fundamental and harmonic that is periodic with depth into the (fixed orientation) crystal. The degree of conversion, in that case, simply depends on the thickness of crystal [15]. But it was known early [13,18] that if the crystal orientation itself (direction of the optical axis) could be periodically changed within the crystal, this mismatch could be partially compensated for. This is the concept of periodic-poling, which can lead to very high harmonic conversion. We expect to see this effect, to some degree, in the guided modes of our 2D model by virtue of the fact that, every time our fundamental wave reflects off of the crystal-vacuum interface, the optical axis essentially changes direction. Strictly speaking, in the case of periodic-poling, the optical axis is rotated 180° in each successive slab of crystal that the beam penetrates; with hexagonal geometry, the axis rotation will be much less and is not meant to compensate for phase mismatch.

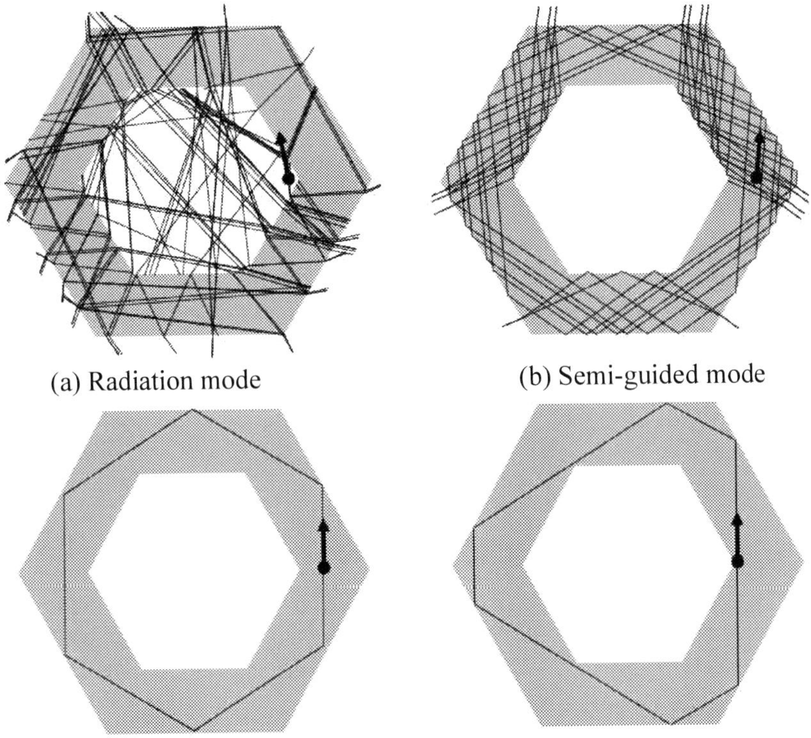

(a) Radiation mode (b) Semi-guided mode

(c) Guided mode, uniform (d) Guided mode, non-uniform

Figure 10. Examples of radiation modes (a,b) and guided modes (c,d) as outputs from the 2D model. The location and direction of the initial ray are shown as an arrow. Rays that escape the crystal are shown as short path segments outside the crystal structure.

The model can be setup such that the initial ray can originate outside the crystal, from within the crystal itself or from within the hollow core. As a check on the model, we show the example of *radiation modes* and *guided modes* in Figure 10. Radiation modes are so named because they cannot contain the input light, which ends up radiating out of the structure eventually. Guided modes, on the other hand, rely on total internal reflection to trap the input light indefinitely (provided that the effective index of refraction in the crystal is greater than that of the vacuum, an easy requirement for most materials) and must originate from within the crystal. We refer to Figure 10b as a semi-guided mode in that all energy from the initial beam will eventually be lost, but there is sufficient total internal reflection to slow the process down compared with Figure 10(a). These figures hint at the wide variety of modes that are possible and the range of unusual geometries that one might expect to find.

As we discovered previously, the four possible phase matching directions are $\theta_{optic} \pm \theta_m$ and $(\theta_{optic} + \pi) \pm \theta_m$. Depending on the origin of the fundamental wave and the geometry of the guided mode, the various ricocheting rays may spend a large portion of their time traveling in one of these four directions. We expect that, similar to a standard rectangular KDP crystal, the degree of harmonic conversion for guided modes will vary periodically with path length and how well the ray segments align with the four phase-matching directions. We would also expect that bulk crystal SHG will be many times more significant than surface generated SHG, which should be limited by total internal reflection in guided modes (although we will not test this here).

4.2. Uniform Guided Mode

We are now ready to begin our main simulations of the guided modes of hexagonal structures. We will assume, henceforward, that $\varphi = 45°$, but we will vary our other parameters depending on the situation.

First let us examine the harmonic generation arising from the guided mode shown in Figure 10c. For this, we assume an incident fundamental power of $P_\omega(0) = 1\text{MW/cm}^2$, and let $P_{2\omega}(0) = 0$. We focus on two different optical axis orientations: one in which two of the six ray segments become perfectly phase matched ($\theta_{optic} = \pi/2 - \theta_m = 48.793°$, since we will be launching the initial ray vertically) and one in which none of the ray segments are perfectly phase matched ($\theta_{optic} = 90°$). The latter is also of interest in that two of the six ray segments are parallel to the optical axis, which should lead to a negligible change of harmonic signal for those paths as a further check on our model. We also include two other optical axis orientations that are nearly phase-matched for comparison. In actuality, the optical axis angle can take on only certain values due to the shape of the unit cell of crystal. But, since our model will accept any value, we have the freedom to experiment. We anticipate an overall dependence on phase-matching angle similar to Figure 11.

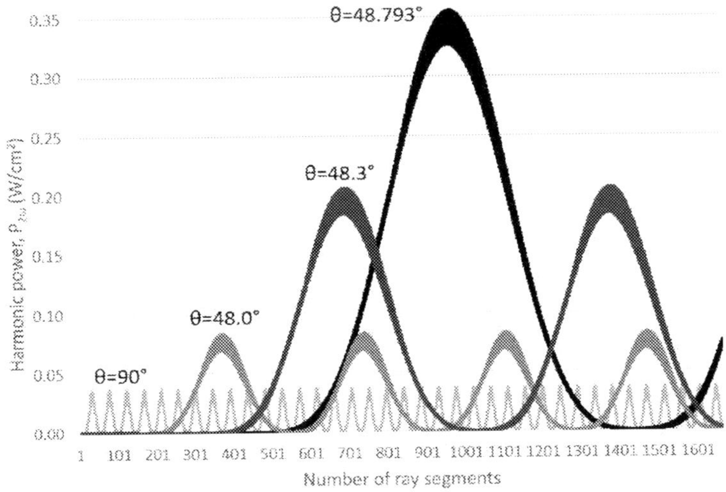

Figure 11. Generation of the second harmonic signal from within the hexagonal crystal (in the uniform guided mode) for various orientations of the optical axis.

We expect that oscillations within the signal should have several different frequency contributions. The guided mode itself is obviously a set of six repeating, 7,500 nm long segments, so we expect oscillation with a period of six units somewhere within the signal, in addition to the more typical SHG spatial period of $2\pi/\Delta k$. It is also worth pointing out that, as we see in eq. (3.6), an amount of $\pi/2$ is added to the accumulated phase every time the beam changes direction upon reflection. We show the values of phase mismatch in Table 1. All of these various periodic components should combine to produce the harmonic signal pattern from our simulation.

We see the results in Figure 11, in which we note the effect of the four different optical axis orientations on the shape and strength of the harmonic signal. All four simulations used the same initial, fundamental power as opposed to the fundamental power being distributed among the four orientations. All signals are periodic with total path length and, in general, the amplitude grows as we turn the optical axis so that phase-matching can occur along at least two of the six ray segments. Not surprisingly, the $\theta_{optic} = 48.793°$ orientation produced the largest peak, with an on-peak conversion efficiency of $\eta = P_{2\omega}/P_{\omega} \approx 10^{-7}$ and a period of 965 ray segments. It arrived at this peak after 970 ray segments which, dividing by six, is approximately 162 complete transits around the hexagon. The $\theta_{optic} = 90°$ orientation produced the smallest peak, with an on-peak conversion efficiency of $\eta \approx 10^{-8}$ and a period of 45 ray segments.

Table 1 – Value of phase mismatch Δkz for different ray segments and optical axis angles when $z = 7{,}500$ nm for each segment in the guided mode pattern.

			Δkz for optical axis angles			
Segment	Length (nm)	Direction	θ = 90°	θ = 48°	θ = 48.3°	θ = 48.793°
1	7500	90°	1.651413	0.051275	0.031855	0.000024
2	7500	150°	1.164712	1.910392	1.917916	1.929888
3	7500	210°	1.164712	1.282755	1.270839	1.250897
4	7500	270°	1.651413	0.051275	0.031855	0.000024
5	7500	330°	1.164712	1.910392	1.917916	1.929888
6	7500	30°	1.164712	1.282755	1.270839	1.250897

We note also that there is some width to each of the traces, particularly at the peaks for the three signals that approach phase-matching, where the harmonic signal rapidly oscillates between two different values. The width is approximately 8% of the peak value. For these three orientations, there is an initial portion of each trace in which there is essentially no harmonic conversion and in which, it seems, the phase differences lead to destructive interference. Just beyond this region, the harmonic signal builds up and becomes perfectly periodic with path length as we would expect.

4.3. Non-Uniform Guided Mode

Next let us examine the harmonic generation arising from the guided mode shown in Figure 10d. We assume the same essential parameters as before, but also note that our six ray segments are no longer uniform: three have lengths of 12,000 nm and three have lengths of 3,000 nm. We can construct a segment-dependent phase mismatch table as before, shown in Table 2.

Table 2. Value for the phase mismatch Δkz for different ray segments and optical axis angles in the non-uniform, guided mode pattern

			Δkz for optical axis angle			
Segment	Length (nm)	Direction	θ = 90°	θ = 48°	θ = 48.3°	θ = 48.793°
1	12000	90°	2.642261	0.082041	0.050968	0.000039
2	3000	150°	0.465885	0.764157	0.767166	0.771955
3	12000	210°	1.863539	2.052408	2.033342	2.001435
4	3000	270°	0.660565	0.020510	0.012742	0.000009
5	12000	330°	1.863539	3.056627	3.068666	3.087820
6	3000	30°	0.465885	0.513102	0.508336	0.500359

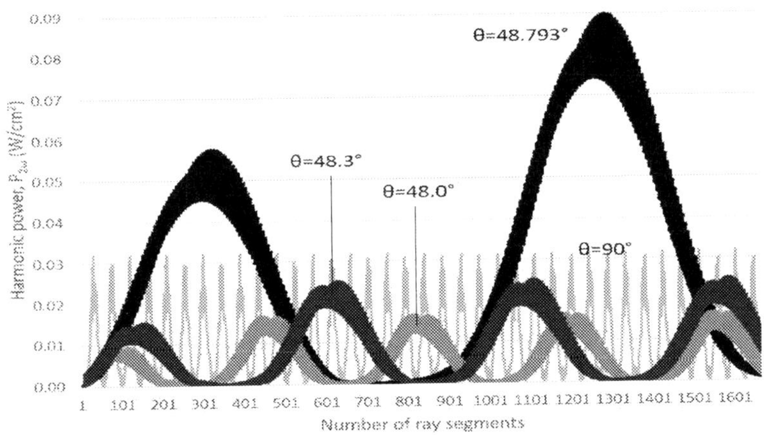

Figure 12. Generation of the second harmonic signal from within the hexagonal crystal (in the non-uniform guided mode) for various orientations of the optical axis.

We see the result in Figure 12, which has some different features compared with the uniform mode results in Figure 12. The harmonic conversion, in general, is less compared with that of the uniform mode. The harmonic signal for $\theta_{optic} = 48.793°$ remains the strongest, with an on-peak conversion efficiency of $\eta = P_{2\omega}/P_{\omega} \approx 10^{-8}$ and the same period of 965 ray segments. The peak was reached after 1290 ray segments which, dividing by six, is 215 complete transits around the hexagon. However, the next largest harmonic signal of the four originated from the $\theta_{optic} = 90°$ orientation with a period of 45 ray segments, indicating greater complexity in the nature of phase mismatch for this geometry. For each of the three orientations that approach phase-matching we see the immediate build-up of an initial, smaller peak that leads, on the next cycle, to a larger peak that sets the height of all successive peaks. We also see a greater width to each trace when, as before, the harmonic signal oscillates rapidly between two values; the width is about 17% of the peak value.

4.4. Comparison of Modes

The signal from the $\theta_{optic} = 90°$ orientation has a much simpler structure than the other orientations, and is easier to compare between the uniform and non-uniform guided modes (see Figure 13 for the comparison). Since these modes had two of the six ray segments aligned to the optical axis, we expected to see no change in the harmonic signal every third segment, leading to the periodic, horizontal portions of the traces interspersed with peaks and troughs. This is evident in the plot. We also see that larger cycle has smaller cycles within it. Each of the larger cycles is identical in structure for the uniform mode, but the non-uniform mode shows more complex structure. In this case, we see the signal only fully repeat after three large cycles due to the smaller cycles operating at a different periodicity.

It is important to point out that the non-uniform mode (as we have defined it) and the corresponding mode created by reflection across the laboratory y-axis represent the end-points of the whole range of possible modes that are totally internally reflected, with each mode having a certain "carrying capacity" of harmonic light as a function of path length. The uniform mode is the middle point between these extremes. Engaging as many of these modes as possible while avoiding destructive interference would make the microcrystalline structure more efficient at generating a harmonic signal compared with employing a single mode. To investigate this carrying capacity, we simulated 52 different guided modes by launching the initial rays vertically with the same y-coordinate but varying the x-coordinate of the launch point in steps of 50 nm. All of these initial rays, which were 100% fundamental power, had a power density of 1 MW/cm^2. In addition, the optical axis was set so that two of the six ray segments of each mode shape were phase-matched. We then found the peak value of the harmonic signal produced for each mode and plotted it. The results are shown in Figure 14.

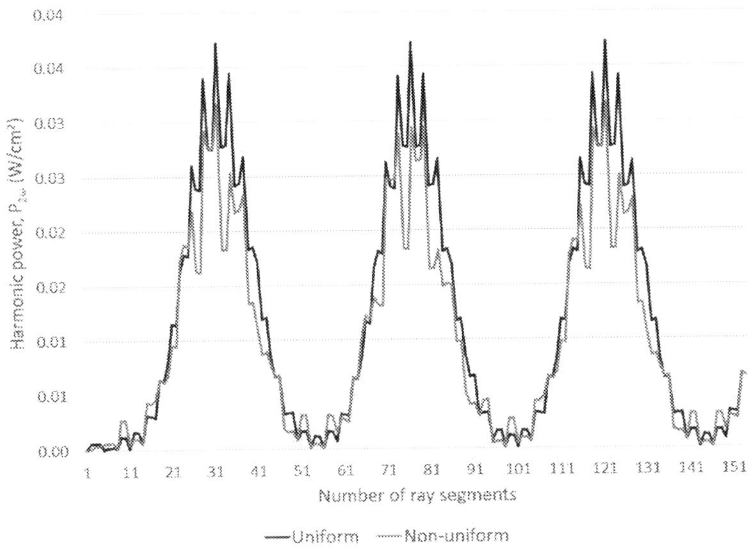

Figure 13. Comparison of the $\theta_{optic} = 90°$ harmonic signals and elongation/ truncation of particular segments for the different mode shapes; note that by segment 145, the uniform mode has completed three identical cycles while the non-uniform mode has fully completed only one due to more complex structure.

For each guided mode, we allowed the simulation to run through 1,670 ray segments, or a total path length of approximately 12.5 mm. Not surprisingly our uniform mode produced the largest peak harmonic power. The peak power then falls smoothly toward a minimum and then, surprisingly, begins to increase as we arrive at the boundary mode shapes. As we noted previously, the peak harmonic signal depends in a complex way on the ray segment lengths and directions since they feed directly into our estimation of the phase mismatch, and it is to be expected that this phase mismatch will have peaks and troughs. The minima correspond with ray segment configuration in which the phase-matched lengths are long compared with the characteristic length, thereby meaning the harmonic reaches an intra-segment peak, and then recedes and strikes the crystal face before it has a chance to rebuild. Since the general shape of the curve in fig 14 is sinusoidal in nature, we can approximate it using a simple model

$$P_{2\omega}(x) \approx P_0 \cos\left[(x - x_0)\frac{\tau}{2\pi}\right] + B_0 \tag{4.1}$$

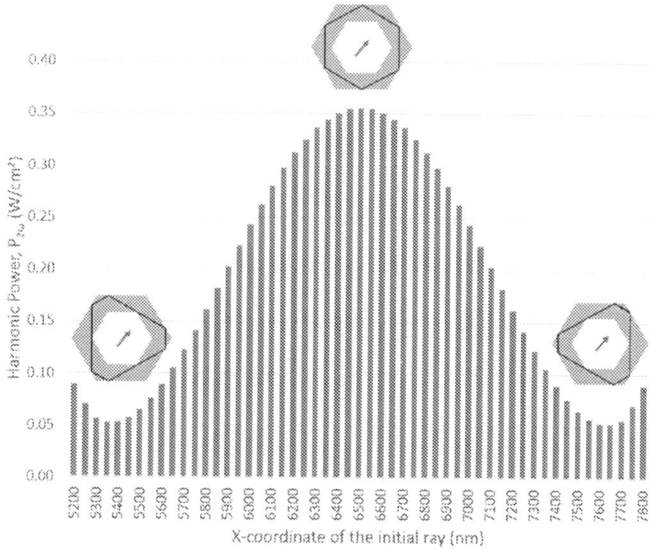

Figure 14. Peak harmonic power as a function of guided mode shape, which is determined by the x-coordinate of the initial ray. Three of the key mode shapes are shown near their respective regions of operation; the internal arrows show the direction of optical axis.

where $x_0 = (D_s/2)\cos 30°$ is the x-coordinate for the vertical ray segment in the uniform guided mode, D_s is the smaller outside diameter of the crystal, B_0 is a baseline harmonic power that will be present for any of the guided modes, P_0 is the maximum power gained above baseline and τ is the period. Integrating over the modes, we estimate a total harmonic signal capacity of 10.089 W/cm^2. Since we input a total of 53 MW/cm^2, this leads to an efficiency of $\eta = P_{2\omega}/P_\omega \approx 10^{-7}$ as we saw previously.

4.5. SHG and Dimensional Scaling

Up to this point we have kept constant the dimensions of the crystal, as shown in Figure 10, since these dimensions were specifically cited in the original research as being a common size among all the fibers grown in a batch. But, smaller and larger fibers were also found. To see what effect the physical size of the crystal has on the production of harmonic power, we varied the overall dimensions of the cross-section of fiber by scaling the lengths involved to maintain the ratio between the inside diameter and the outside diameter. Using the uniform guided mode in every case, with two out of six segments phase-matched, we examined the harmonic power output. We already know that the overall path length contributes to the phase mismatch and, since scaling the dimensions will obviously increase/decrease path length, we expect the peak the shift accordingly.

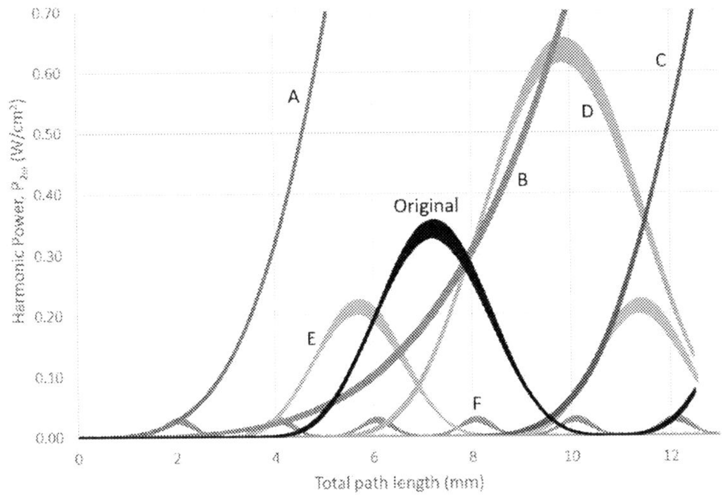

Figure 15. Harmonic power as a function of path length for the uniform mode in variously sized crystals; all signals had the same general sinusoidal shape, the view is restricted to emphasize the variety of peak signals available.

Our results are shown in Figure 15 and summarized in Table 3. We never had a reason to expect that our crystal dimensions were optimized for maximal harmonic signal output, and our data demonstrates this fact. The curve labelled "original" in Figure 15 is the harmonic signal we have been analyzing up until now, which was shown as the largest peak in Figure 12. We adjusted the outside diameter of the structure (the smaller of the two ways to measure it) in discrete steps and re-ran the simulation each time to produce a trace of the output (or harmonic power vs. total path length travelled) and a peak value. In every case, the input fundamental power was identical at 1 MW/cm^2. From Table 3, we see that an outside diameter of 14,800 nm yielded the highest peak value recorded for the harmonic signal at 50.392 W/cm^2, indicating a conversion efficiency of $\eta = P_{2\omega}/P_{\omega} \approx 10^{-5}$ or two orders of magnitude larger than what was possible with our original dimensions, and only one order of magnitude less than the high efficiency value of $\eta \approx 10^{-4}$ originally witnessed in the experimental fibers [1]. It is expected, although we did not test it here, that the harmonic signal is periodic with the cross-sectional diameter of crystal by virtue of the path length of the totally internally reflected beam. This result shows at least, in principle, that this geometry could be used for high efficiency harmonic conversion even in ordinary KDP crystals *provided* that they could be manufactured with the desired dimensions and orientation of the optical axis. The new experimental fibers have these properties naturally, and their conversion efficiency is more impressive given that they include losses while our model does not.

Table 3. Peak harmonic power output as a function of crystal size; smaller outside diameter (OD) refers to the distance across the hexagon mid-side. Labels correspond with those in Figure 15. A full transit around the hexagonal structure is six ray segment lengths

Label	Smaller OD (nm)	Ray Segment (nm)	Harmonic Power (W/cm^2)
A	14,750	7,375	2.803
B	14,800	7,400	50.392
C	14,900	7,450	1.620
D	14,950	7,475	0.655
Original	15,000	7,500	0.356
E	15,050	7,525	0.225
F	15,500	7,750	0.032

All of our modeling work so far has been strictly 2D in that we have simulated light rays bouncing around in the cross-section of the crystal. But, in reality, these fibers are 3D structures much longer than they are wide and light propagation down the long axis has its own dynamics. We turn now to examination of the model from a 3D perspective and what we might expect from bundles of fibers acting together.

5. Extension to 3D and Multi-Fiber Bundles

5.1. Incorporating Input Angle with Phase Matching

As was previously mentioned, our numerical model described the ray optics in a flat plane cut perpendicular to the long axis (crystallographic *b*-axis) of the structure. It was also noted that guided modes require the initial ray to launch from within the crystal, which can be implemented with end-coupling the pump source to the microcrystalline fiber. The research suggests that part of the reason for such high harmonic conversion. That is, the angular distribution of the input pump light ensures that a significant portion of the fundamental mode is traveling constantly in the phase-matching direction even as it ricochets off the vacuum-crystal interfaces. While this is difficult to visualize and relies on vector decomposition, it can be easily shown that such a situation exists and certainly enhances SHG for guided modes.

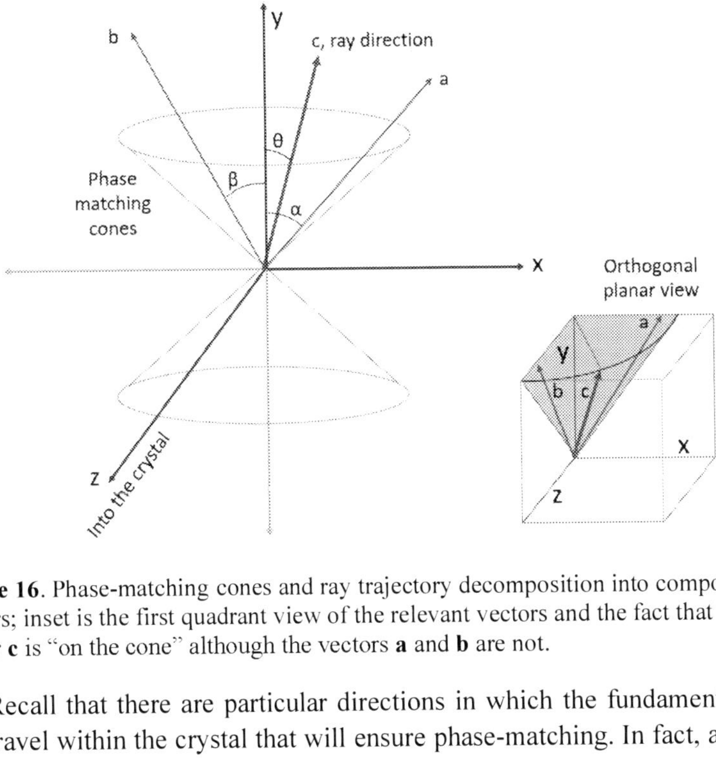

Figure 16. Phase-matching cones and ray trajectory decomposition into component vectors; inset is the first quadrant view of the relevant vectors and the fact that the vector **c** is "on the cone" although the vectors **a** and **b** are not.

Recall that there are particular directions in which the fundamental ray can travel within the crystal that will ensure phase-matching. In fact, as long as the ray travels along the surface of one of two cones with respect to the optical axis, phase-matching is assured. The challenge then is to find the combination of input angle and 2D ray angle that will yield the requisite $\theta_m = 41.207^c$ and keep our ray "on the cone," so to speak. This should take the form of a 3D, trigonometric constraint.

See Figure 16 for our general setup. There is quite a bit going on visually in this diagram; so, we will address each piece separately. We take a microscopic view and imagine that the crystal fills half of all space $(z > 0)$

and our coordinate origin lies on the vacuum-crystal interface. As usual, we have a right-handed coordinate system in which the x- and y- axes have the same orientation as before in our 2D model, and we now include the z-axis to indicate the long axis of the crystalline structure. The twin-cone diagram is superimposed on the coordinate system to indicate that we have chosen the optical axis to be, for the sake of simplicity, the y-axis (or $\theta_{optic} = 90°$ in our original system). In our old 2D model, we measured the angle of a ray's trajectory with respect to the positive x-axis; here we will call it α and measure it with respect to the positive y-axis to simplify the formulas that follow. The ray trajectory is the vector labelled **a** in the diagram and lies in the x-y plane. We also plan to launch the ray (at least partially) in the z-direction. We call that angle β and measure it from the y-axis toward the z-axis; the launch trajectory vector is labelled **b** in the diagram and lies in the y-z plane.
The question then becomes – if a 3D ray vector **c** has components **a** and **b**, will it lie on either of the two phase-matching cones? Stated more explicitly, do α and β have values such that the resulting 3D angle θ (the angle that **c** makes with the y-axis) has the right value for phase-matching? We start with the formula we want to satisfy, which is $\tan\theta = r/y$, where $r^2 = x^2 + z^2$. We can also use $\tan\alpha = x/y$ and $\tan\beta = z/y$ to arrive at the general relation

$$tan^2(\theta) = tan^2(\alpha) + tan^2(\beta) \tag{5.1}$$

As we can see from the above relation, there is no explicit reference to the optical axis angle and, as such, the angles are relative, and can be applied to any crystal orientation since we can realign our coordinate system to whatever orientation the optical axis may have. In order to get the most harmonic conversion possible, we want the fundamental wave to spend as much time as possible moving inside the crystal in a phase-matching direction. As such, we want to maximize the overall path length by using a very shallow launch angle, e.g., $\beta < 5°$, so that ideally the path of the ray would resemble a tightly-wound, hexagonal-sided helix in a guided mode. Experimentally this would translate to placing the minimum beam waist of the pump laser on one end of the hexagonal crystalline structure such that the beam path is almost perpendicular to the b-axis. This is similar to grazing incidence and, while only a small fraction of the pump light will couple to the crystal, it will enter at such a small angle β that all the captured rays will circulate many times before reaching the other end of fiber.

The additional dimension provides a measure of freedom in ray propagation angle that we did not have with the 2D model. In that model, we were limited to four discrete directions to achieve perfect phase-matching, or two lines on each of the twin cones. A third dimension provides us with the ability to phase-match on half of the upper cone and half of the lower cone, or a (limited) continuum of directions directed into the crystal. This, in turn, implies that our fundamental rays, when entering the crystal, could have an angular distribution that ensures phase-matching in more directions and modes. Let us examine the case of nonzero β and the effect that it has on the harmonic signal. If the distance of a ray segment in the 2D uniform, guided mode is s, the 3D distance traveled through the crystal with an input angle β is:

$$d = s/\cos\beta \tag{5.2}$$

We would use this distance d rather than z in previous equations, such as eq. (3.1), to adapt them for three dimensions. For clarity, the distance traveled along the long axis of the crystal would be:

$$z = s/\sin\beta \tag{5.3}$$

with the resulting Pythagorean relationship:

$$d^2 = s^2 + z^2 \tag{5.4}$$

Since the harmonic signal generated, as well as the phase mismatch accumulated, depends heavily on the distance travelled through the crystal, a nonzero input angle (which increases that distance) should have a similar effect as increasing the dimensions of the hexagonal crystal, which we explored in the previous chapter. With $\beta = \theta_m$, which is the largest value allowed for the input angle to ensure any degree of phase matching, from eq. (5.2), we would have $d \approx 1.33s$, indicating a 33% increase in ray segment length. But, as we see from eq. (3.4), greater distance alone does not lead to an increase in the harmonic signal. Thus far we have used this expression for the wave vector phase mismatch:

$$\Delta k = \frac{2\omega_1}{c}\left[n_{eff}(\theta, \lambda_2) - n_o(\lambda_1)\right] \tag{5.5}$$

But, now we need to express θ in terms of β and α using eq. (5.1) in order to generalize it for use in 3D:

$$\theta \to \tan^{-1}\left(\sqrt{\tan^2(\alpha) + \tan^2(\beta)}\right) \tag{5.6}$$

Indeed, using $\beta = \theta_m$ makes it impossible to keep more than two ray segments of the circulating mode "on the cone." Each possible value of β would require its own value for the optical axis angle, in order to perfectly phase match two of the six segments in the uniform guided mode; meaning that a distribution of β would require a distribution of θ_{optic} which, of course, is not feasible in a single grown crystal structure in which the optical axis has a single direction relative to the crystallographic axes. At any rate, having the option to phase match using a combination of α and β allows us a small measure of freedom that we did not have in the strictly 2D case.

5.2. Multi-Fiber Bundles and SHG

The researchers also discussed the interesting possibility of creating a bundle [17] of these hexagonal fibers which, acting as an array of crystals, could have interesting properties. The resulting microarray would resemble a photonic crystal fiber in which a regular arrangement of holes and crystal material are placed at intervals to effectively trap and conduct input light. The dimensions of fibers obtained experimentally are approximately the right size to grow such a photonic fiber naturally rather than starting with a bulk crystal and engineering the necessary gaps to create the desired pattern. Shown in Figure 17 is an example of such a bundle in which the individual crystal fibers are joined together along their external, longitudinal faces. We show each fiber as having a different orientation of its optical axis since, in practice, it would be difficult to align them all when assembling the bundle. Another parameter to consider is whether they can be bonded together such that the crystal-crystal interfaces are refraction-less, and a ray traveling toward the boundary would pass through it unaffected. While this is unlikely, it presents the possibility that the hexagonal cavities with vacuum-crystal interfaces would provide the chance for total internal reflection; alternatively, any light that enters those cavities would eventually escape by virtue of radiation modes. It should be pointed out that, even Figure 17 is an idealization, given that the individual fibers naturally form in a wide range of diameters and lengths, leading to a random assortment of fibers that might be bound into awkward, asymmetric bundles.

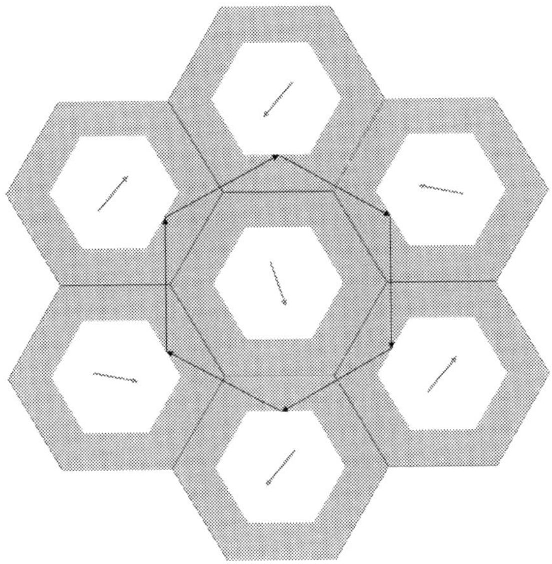

Figure 17. End view of a crystal fiber bundle assembled from several individual fibers. Here we assume a random arrangement of the optical axes among the fibers (center of each hexagon) and an example uniform guided mode (connected rays circling the center of the bundle) that uses the gaps for total internal reflection.

Imagining for a moment that we did have the freedom to design these bundles, there are many degrees of freedom that we might exploit, in order to amplify the harmonic signal generated. We saw previously that overall fiber size is crucial to facilitating harmonic generation, and so, we would choose to create our bundle using size-optimized fibers. If we had the freedom to set the optical axis orientation for each constituent fiber in the bundle, it would be possible to ensure that all six ray segments of the mode shown in Figure 17 could have half of their length perfectly phase-matched. In essence, this would translate to phase-matching half (3/6 segments) of the uniform mode whereas a single fiber can only phase-match one-third (2/6 segments) of the mode, which would also increase the harmonic signal. At any rate, the ability to scale and leverage the basic hexagonal structure of these fibers will be a promising area of research for SHG.

Conclusion

Research continues on these microcrystalline KDP fibers and efficient SHG has been observed in fibers with increasingly smaller diameters. Where harmonic light was once an oddity that was quite difficult to demonstrate, these fibers make it possible to incorporate optical harmonics in ever-smaller application packages. It is the hope that the simple, low-resource model described here can help novice researchers in the field understand SHG in such a specialized context.

Acknowledgements

Many thanks to my friend, Santos, whose physics discussions are always enlightening, and to my mentor, Chris, for his professional and personal guidance.

References

[1] Kuzyk, M.G. *"Nonlinear Optics: A Student's Perspective"*, Create Space Independent Publishing Platform, North Charleston, SC (2017).

[2] Besse, C., B. Bidégaray-Fesquet, A. Bourgeade, P. Degond, and O. Saut, "A Maxwell-Bloch model with discrete symmetries for wave propagation in nonlinear crystals: an application to KDP", *ESAIM: Math. Model. Numer. Anal.*, Vol. 38, No. 2, pp.321-344, 2004.

[3] Tien, P.K. "Light waves in thin films and integrated optics", *Appl. Opt.*, vol. 10, no.11, pp. 2395–2413, 1971.

[4] Zemon, S., R.R. Alfano, S.L. Shapiro, and E. Conwell, "High-power effects in nonlinear optical waveguides", *Appl. Phys. Lett.*, vol. 21, no. 7, pp. 327-329, 1972.

[5] E. Conwell, "Theory of second-harmonic generation in optical waveguides", *IEEE J. Quantum Electron.*, vol. 9, no. 9, pp.867-879, 1973.

[6] Lamoth, Andre. *"Tricks of the Windows Game Programming Gurus"*, 2nd Ed., Sams, Indianapolis, Indiana, USA (2002).

[7] Shen, Y.R. *"The Principles of Nonlinear Optics"*, Wiley, New York, NY (1984).

[8] Bauer, L. and E.L. Reiss, "Cutoff wavenumbers and modes of hexagonal waveguides", *SIAM J. Appl. Math.*, vol. 35, no.3, pp. 508–514, 1978.

[9] Komarov, V.V. "Eigenmodes of regular polygonal waveguides", *J. Infrared, Millim. and Terahertz Waves*, vol. 32, no. 1, pp. 40–46, 2011.

[10] Mizrahi, V., U. Österberg, J.E. Sipe, and G.I. Stegeman, "Test of a model of efficient second-harmonic generation in glass optical fibers", *Opt. Lett.*, vol. 13, no. 4, pp. 279–281, 1988.

[11] Aghaie, K.Z., V. Dangui, M.J.F Digonnet, S. Fan, and G.S. Kino, "Classification of the core modes of hollow-core photonic-bandgap fibers", *IEEE J. Quantum Electron.*, vol. 45, no. 9, pp.1192-1200, 2009.

[12] Armstrong, J.A., N. Bloembergen, J. Ducuing, and P.S. Pershan, "Interactions between light waves in a nonlinear dielectric", *Phys. Rev.*, vol. 127, no. 6, pp. 1918-1939, 1962.

[13] Boyd, R.W. "*Nonlinear Optics*", 2nd Ed., Academic Press, an Imprint of Elsevier Science, New York, NY (2003).

[14] Bloembergen, N. "*Frontiers in Physics: Nonlinear Optics*", A Lecture Note and Reprint Series, W.A. Benjamin, Inc. Publishers, New York, NY (1965).

[15] Maker, P.D., R.W. Terhune, M. Nisenoff, and C.M. Savage, "Effects of dispersion and focusing on the production of optical harmonics", *Phys. Rev. Lett.*, vol. 8, no. 1, pp. 21-23, 1962.

[16] Jerphagnon, J. and S. K. Kurtz, "Maker fringes: a detailed comparison of theory and experiment for isotropic and uniaxial crystals", *J. Appl. Phys.*, vol. 41, no. 4, pp. 1667-1681, 1970.

[17] Ren, Y., X. Zhao, E.W. Hagley, and L. Deng, "Ambient-condition growth of high-pressure phase centrosymmetric crystalline KDP microstructures for optical second harmonic generation", *Sci. Adv.*, vol. 2, no. 8, 2016.

[18] Zernike, F. and J.E. Midwinter, "*Applied Nonlinear Optics*" reprint (2006) of John Wiley & Sons, Inc. New York, NY (1973).

Additional Resources

Sipe, J.E. "New Green-function formalism for surface optics", *J. Opt. Soc. Am B*, vol. 4, no. 4, pp. 481–489, 1987.

Nelmes, R.J., Z. Tun, and W.F. Kuhs, "A compilation of accurate structural parameters for KDP and DKDP, and a users' guide to their crystal structures", *Ferroelectrics*, vol. 71, no. 1, pp.125-141, 1987.

Shen, Y.R. "Optical second harmonic generation at interfaces", *Annu. Rev. Phys. Chem.*, vol. 40, no. 1, pp. 327–50, 1989.

Heinz, T.F. "Second-Order Nonlinear Optical Effects at Surfaces and Interfaces", In *Nonlinear Surface Electromagnetic Phenomena*, H.E. Ponath and G.I. Stegeman (Eds.), Elsevier Science Publishers, New York, NY, USA, 1991.

Chapter 4

Design and Analysis of Photonic Crystal-Based Integrated Optical Devices Using the Finite Difference Method

Mohsen Choubani*, **Hassen Maaref and Faouzi Saidi**
Laboratoire de Micro-Optoélectronique et Nanostructures (LR99ES29), Département de Physique, Université de Monastir, Tunisia

Abstract

The finite-difference time-domain (FDTD) method was presented for the analysis of several practical situations such as frequency selective surfaces (FSS), electromagnetic bandgap (EBG), optical filters, waveguides, and demultiplexer based on photonic crystals. This was done along with the implementation of an artificial absorbing boundary medium known as the perfectly matched layer (PML). This medium enables us to limit the computational domain required for simulation. Using Gaussian pulse excitation, the reflection, transmission and absorption coefficients, and field distributions may be computed inside and outside of the structures in one simulation. Also, the FDTD formulation may be developed for any polarization: transverse electric (TE) or transverse magnetic (TM) fields at any frequency of operation (microwave, millimeter-wave, optical), any number of dielectric layers with arbitrary thicknesses, and any inhomogeneous dielectric profile.

Keywords: demultiplexer, electromagnetic bandgap, FDTD; frequency selective surfaces, inhomogeneous media, optical filters, perfectly matched layers, photonic crystals

---

* Corresponding Author's Email: mohsenchoubani3@yahoo.fr.

In: Second Harmonic Generation
Editors: Abdel-Baset M. A. Ibrahim and Pankaj Kumar Choudhury
ISBN: 978-1-68507-888-1

1. Introduction

To overcome the limitations of analytical methods, numerical methods have been developed. The commonly used methods are the finite element method (FEM) [1], method of moment (MOM) [2], and finite difference time domain (FDTD) method [3, 4]. Among these, the FDTD method has been proved to be one of the most effective numerical methods in the study of complex structures. As a direct solution to Maxwell's equations, FDTD offers a simple yet straightforward way to model complex structures. Since it is a time-domain solver, it is convenient in dealing with the characteristics of any structure over a wide frequency band.

The foundation of FDTD was laid down by Yee in 1966 [3]. Yee chose a geometric relationship for the spatial sampling of vector components of the electric and magnetic fields that enables representing both the differential and integral forms of Maxwell's equations in a robust manner. The FDTD algorithm, as proposed by Yee in his original paper, is second-order accurate in both time and space. Furthermore, in this algorithm, the numerical dispersion effects can be kept small by using a cell size that is sufficiently small in comparison to the wavelength, say on the order of $\lambda/20$ or $\lambda/30$.

The use of the FDTD method provides a rigorous solution to Maxwell's equations and does not have any approximations or theoretical restrictions. Since the first algorithm, written by Yee in 1966 [3], the FDTD method has emerged as a primary means to computationally model many scientific and engineering problems dealing with the electromagnetic wave interactions with material structures. Moreover, the FDTD method is an application of the finite difference scheme (commonly used in solving differential equations) to solve Maxwell's equations. In FDTD, the space is divided into small portions called cells. On the surfaces of each cell, there are assigned points. Each point in the cell is required to satisfy Maxwell's equations. With the appropriate choice of the points at which the various field components are to be evaluated, the set of finite difference equations can be solved and the solution will satisfy the boundary. In this way, electromagnetic waves are simulated to propagate in a numerical space, almost as they do in the real physical world.

As well, the FDTD method is a widely used technique for solving electromagnetic problems with many advantages, such as – the method is a direct solution of Maxwell's time-domain equations, and consequently, it is a complete full-wave solution that contains no approximations; it is extremely general in the materials and geometries that it can analyze; and, finally, the memory requirements of FDTD are significantly less than the other methods, viz. the plane wave method (PWM).

The principal disadvantage of the classical implementation of FDTD is that all structures must conform to a Cartesian grid. Consequently, all curved surfaces must be modeled by a *stair-step* approximation, which may introduce errors in the results.

FDTD is good for radio frequency (RF) and microwave applications. It is not too difficult to write computer programs to implement FDTD and there is no need to solve matrix equations. The final results of FDTD computer programs are displayed in movies, and therefore, are easy to understand and good for physical insight. As a result, FDTD is suitable for teaching purposes.

In this chapter, an introduction to Maxwell's equations and the FDTD method is presented in Section 2. The numerical formulations for the absorbing boundary conditions, such as the perfectly matched layer (PML), are formulated [5]. Also, the analysis of stability and numerical dispersion are studied. Moreover, the choice of source excitation is well discussed in this section. In Section 3, numerical examples are simulated and compared to those using Hill's method [6]. Furthermore, various applications are discussed, such as the problem of modeling the frequency selective surface (FSS) structures, electromagnetic bandgap (EBG) structures, optical filters, waveguides, and demultiplexer based on photonic crystals.

2. Formulations of The Yee's FDTD Algorithm

The Yee algorithm simultaneously deals with both the electric and magnetic fields in time and space using the coupled form of Maxwell's curl equations, rather than by solving the wave equation for either the electric field or the magnetic field alone.

2.1. Maxwell's Equations

The algorithm of the FDTD method is based on the differential form of Maxwell's equations [7], given as

$$\vec{\nabla} \wedge \vec{E} = -\frac{\partial \vec{B}}{\partial t} - \sigma_m \vec{H} \tag{1}$$

$$\vec{\nabla} \wedge \vec{H} = \sigma_e \vec{E} + \frac{\partial \vec{D}}{\partial t} \tag{2}$$

$$\vec{\nabla} \cdot \vec{D} = \rho_e \tag{3}$$

$$\vec{\nabla} \cdot \vec{B} = 0 \tag{4}$$

where the constitutive relations are:

$$\vec{D} = \varepsilon \cdot \vec{E} \tag{5}$$

$$\vec{B} = \mu \cdot \vec{H} \tag{6}$$

Here $\vec{D}$ is the electric displacement field vector in the units of C.m^{-2}, $\vec{B}$ is the magnetic flux (T), $\vec{E}$ is the electric field (V.m^{-1}), $\vec{H}$ is the magnetic field (A.m^{-1}), σ_m is the Magnetic conductivity (Ω.m^{-1}), σ_e is the electrical conductivity (S.m^{-1}), ε is dielectric permittivity (F.m^{-1}), and μ is the magnetic permeability (H.m^{-1}).

Let us consider a region of space that has no electric or magnetic current sources. Substituting Eqs. (5) and (6) into Eqs. (1) and (2), we obtain the following system of six coupled scalar equations under the Cartesian coordinates:

$$\frac{\partial H_x}{\partial t} = -\frac{1}{\mu}\left[\frac{\partial E_z}{\partial y} - \frac{\partial E_y}{\partial z} + \sigma_m H_x\right] \tag{7}$$

$$\frac{\partial H_y}{\partial t} = -\frac{1}{\mu}\left[\frac{\partial E_x}{\partial z} - \frac{\partial E_z}{\partial x} + \sigma_m H_y\right] \tag{8}$$

$$\frac{\partial H_z}{\partial t} = -\frac{1}{\mu}\left[\frac{\partial E_y}{\partial x} - \frac{\partial E_x}{\partial y} + \sigma_m H_z\right] \tag{9}$$

$$\frac{\partial E_x}{\partial t} = \frac{1}{\varepsilon}\left[\frac{\partial H_z}{\partial y} - \frac{\partial H_y}{\partial z} - \sigma_e E_x\right] \tag{10}$$

$$\frac{\partial E_y}{\partial t} = \frac{1}{\varepsilon}\left[\frac{\partial H_x}{\partial z} - \frac{\partial H_z}{\partial x} - \sigma_e E_y\right] \tag{11}$$

$$\frac{\partial E_z}{\partial t} = \frac{1}{\varepsilon}\left[\frac{\partial H_y}{\partial x} - \frac{\partial H_x}{\partial y} - \sigma_e E_z\right] \tag{12}$$

The partial differential Eqs. (7)–(12) form the basis of the FDTD numerical algorithm for modeling electromagnetic wave interactions with arbitrary three-dimensional objects. Yee's FDTD scheme discretizes Maxwell's curl equations by approximating the time and space first-order partial derivatives with central differences and then solving the resulting equations by using a leapfrog scheme.

3. Yee's FDTD Scheme

3.1. The Yee Cell

In this section, we describe the notation introduced by Yee and the finite difference approximations used to numerically evaluate the system of analytical Eqs. (7)–(12). Using Yee's notations, a space point inside a uniform rectangular lattice grid is denoted as $(i, j, k) = (i\Delta x, j\Delta y, k\Delta z)$. Here, Δx, Δy, and Δz are, respectively, the lattice space increments (in meters) in the x-, y- and z-directions ($\Delta z = \Delta y = \Delta x$, if the unit cell is cubic), while i, j, and k are integers multiplying the corresponding increments. Using this notation, we denote any function of space and time, the electric or the magnetic field, for instance, evaluated at a discrete point in the grid as

$$U(i\Delta x, j\Delta y, k\Delta z, n\Delta t) = U^n(i, j, k).$$

where $n\Delta t$ is the time increment in seconds – commonly known as the time-step; n is the time integer and U denotes any of the E- or H-field components.

To evaluate numerically the space and time derivatives in Eqs. (7)–(12), Yee used the second-order accurate centered finite-difference approximation. To illustrate the approximation of derivatives with a finite difference, we consider the function $f(x)$ shown in Figure 1. Suppose that we want to approximate the first derivative of this function about the point x_0. A more general way to derive the finite difference formulas is to use a Taylor series about x_0. We recall the Taylor series approximations for $f(x_0+\Delta x/2)$ and $f(x_0-\Delta x/2)$:

$$f(x_0+\frac{\Delta x}{2})=f(x_0)+\frac{\Delta x}{2}f'(x_0)+0(\Delta x^2) \tag{13}$$

$$f(x_0-\frac{\Delta x}{2})=f(x_0)-\frac{\Delta x}{2}f'(x_0)+0(\Delta x^2) \tag{14}$$

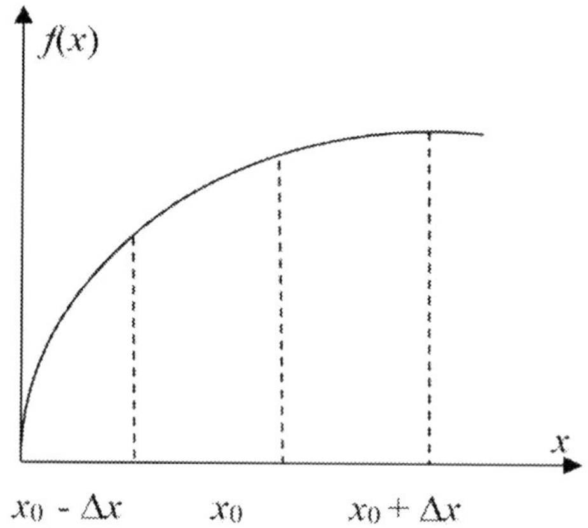

Figure 1. Approximation of the derivative of a function $f(x)$ with a finite difference.

Subtracting the above two Taylor series formulas, we get

$$f'(x_0)=\frac{f(x_0+\frac{\Delta x}{2})-f(x_0-\frac{\Delta x}{2})}{\Delta x}+0(\Delta x^2) \tag{15}$$

where $0(\Delta x^2)$ means that the error in the above equation is of the order of (Δx^2). The concept of the Yee cell (Figure 2) shows us the sampling points of the field variables $\vec{H}$ and $\vec{E}$. Each field variable is surrounded by four fields of the other types. According to Maxwell's equations, the time derivative of a field variable is related to the variable itself and the space derivative of another field(s) in space (Figure 3).

It is seen in Eq. (9) that $\frac{\partial H_z}{\partial t}$ is dependent on itself by the term $-\frac{\sigma_m H_z}{\mu}$, to the derivative of E_x in y dimension by the term $\frac{1}{\mu}\frac{\partial E_x}{\partial y}$, and to the derivative of E_y in x dimension by the term $-\frac{1}{\mu}\frac{\partial E_y}{\partial x}$. However, H_z component is sampled at points $((i + 1/2).\Delta x, (j + 1/2).\Delta y, k.\Delta z)$, the field variable E_x is sampled at $((i + 1/2).\Delta x, j.\Delta y, k.\Delta z)$, and E_y at $((i + 1).\Delta x, (j + 1/2).\Delta y, k.\Delta z)$. In Eq. (9), to calculate the value of $\frac{\partial H_z}{\partial t}$ at location $((i + 1/2).\Delta x, (j + 1/2).\Delta y, k.\Delta z)$, we also need the values of $\frac{\partial E_x}{\partial y}$ and $\frac{\partial E_y}{\partial x}$ at the same point. According to the Yee algorithm and using Eq. (15), we have:

$$\frac{\partial E_x(i+\frac{1}{2}, j+\frac{1}{2}, k)}{\partial y} = \frac{E_x(i+\frac{1}{2}, j+1, k) - E_x(i+\frac{1}{2}, j, k)}{\Delta y} \tag{16}$$

$$\frac{\partial E_y(i+\frac{1}{2}, j+\frac{1}{2}, k)}{\partial x} = \frac{E_y(i+1, j+\frac{1}{2}, k) - E_y(i, j+\frac{1}{2}, k)}{\Delta x} \tag{17}$$

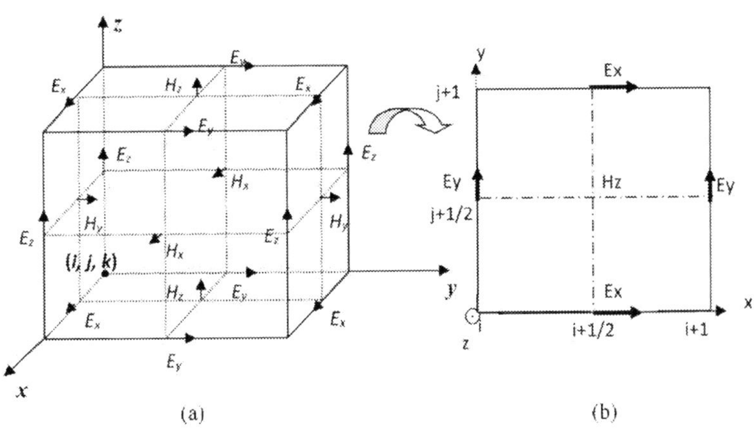

Figure 2. Yee cell for (a) 3D case, and (b) 2D case to evaluate $H_z^{n+1/2}(i+1/2, j+1/2, k)$.

The Yee cell concept shows us how to discretize derivatives of fields in space. Now, and according to the leapfrog processing, we will present and see how the derivatives of field variables in time are discretized.

3.2. The Leapfrog Processing Scheme

Both the $\vec{H}$ and $\vec{E}$ fields are updated within a one-time step Δt. However, they are not sampled at the same moment. In Yee's algorithm, if the $\vec{H}$ fields are updated at time $n.\Delta t$, the $\vec{E}$ fields are sampled at time $(n+1/2).\Delta t$. If we write the time explicitly, Eq. (9) becomes:

$$\frac{\partial H_z^n}{\partial t} = -\frac{1}{\mu}\left[\frac{\partial E_y^n}{\partial x} - \frac{\partial E_x^n}{\partial y} + \sigma_m H^n{}_z\right] \tag{18}$$

where:

$$\frac{\partial H_z^n(i+\frac{1}{2},j+\frac{1}{2},k)}{\partial t} = \frac{H_z^{n+\frac{1}{2}}(i+\frac{1}{2},j+\frac{1}{2},k) - H_z^{n-\frac{1}{2}}(i+\frac{1}{2},j+\frac{1}{2},k)}{\Delta t} \tag{19}$$

and:

$$H_z^n(i+\frac{1}{2},j+\frac{1}{2},k) = \frac{H_z^{n+\frac{1}{2}}(i+\frac{1}{2},j+\frac{1}{2},k) + H_z^{n-\frac{1}{2}}(i+\frac{1}{2},j+\frac{1}{2},k)}{2} \tag{20}$$

Now, by combining the two equations brought by the interleaved coordinate concept (Eqs. (16)–(17) and the two equations brought by leapfrog scheme (Eqs. (19)–(20)), we can discretize Eq. (18) as follows:

$$\frac{H_z^{n+\frac{1}{2}}(i+\frac{1}{2},j+\frac{1}{2},k)-H_z^{n-\frac{1}{2}}(i+\frac{1}{2},j+\frac{1}{2},k)}{\Delta t}=$$
$$-\frac{1}{\mu(i+\frac{1}{2},j+\frac{1}{2},k)}\left\{\begin{array}{l}\left[\left[\dfrac{E_y^n(i+1,j+\frac{1}{2},k)-E_y^n(i,j+\frac{1}{2},k)}{\Delta x}\right]-\left[\dfrac{E_x^n(i+\frac{1}{2},j+1,k)-E_x^n(i+\frac{1}{2},j,k)}{\Delta y}\right]\right]\\ +\sigma_m(i+\frac{1}{2},j+\frac{1}{2},k)\left[\dfrac{H_z^{n+\frac{1}{2}}(i+\frac{1}{2},j+\frac{1}{2},k)+H_z^{n-\frac{1}{2}}(i+\frac{1}{2},j+\frac{1}{2},k)}{2}\right]\end{array}\right\} \tag{21}$$

Rearranging the terms, the above equation simplifies to:

$$H_z^{n+1/2}(i+1/2,j+1/2,k)=\left(\frac{1-\dfrac{\sigma_m(i+1/2,j+1/2,k)\Delta t}{2\mu(i+1/2,j+1/2,k)}}{1+\dfrac{\sigma_m(i+1/2,j+1/2,k)\Delta t}{2\mu(i+1/2,j+1/2,k)}}\right)H_z^{n-1/2}(i+1/2,j+1/2,k)$$
$$-\left[\frac{\Delta t}{\mu(i+1/2,j+1/2,k)}\Big/1+\frac{\sigma_m(i+1/2,j+1/2,k)\Delta t}{2\mu(i+1/2,j+1/2,k)}\right]$$
$$\times\left[\frac{E_y^n(i+1,j+1/2,k)-E_y^n(i,j+1/2,k)}{\Delta x}-\frac{E_x^n(i+1/2,j+1,k)-E_x^n(i+1/2,j,k)}{\Delta y}\right] \tag{22}$$

Similarly, Eqs. (7)–(12) are discretized as:

$$H_x^{n+1/2}(i,j+1/2,k+1/2)=\left(\frac{1-\dfrac{\sigma_m(i,j+1/2,k+1/2)\Delta t}{2\mu(i,j+1/2,k+1/2)}}{1+\dfrac{\sigma_m(i,j+1/2,k+1/2)\Delta t}{2\mu(i,j+1/2,k+1/2)}}\right)H_x^{n-1/2}(i,j+1/2,k+1/2)$$
$$-\frac{\dfrac{\Delta t}{\mu(i,j+1/2,k+1/2)}}{1+\dfrac{\sigma_m(i,j+1/2,k+1/2)\Delta t}{2\mu(i,j+1/2,k+1/2)}}\left[\begin{array}{l}\dfrac{E_z^n(i,j+1,k+1/2)-E_z^n\left(i,j,k+\frac{1}{2}\right)}{\Delta y}\\ -\dfrac{E_y^n(i,j+1/2,k+1)-E_y^n(i,j+1/2,k)}{\Delta z}\end{array}\right] \tag{23}$$

$$H_y^{n+1/2}\left(i+1/2,j,k+\frac{1}{2}\right)=\left(\frac{1-\dfrac{\sigma_m\left(i+1/2,j,k+\frac{1}{2}\right)\Delta t}{2\mu\left(i+1/2,j,k+\frac{1}{2}\right)}}{1+\dfrac{\sigma_m\left(i+1/2,j,k+\frac{1}{2}\right)\Delta t}{2\mu\left(i+1/2,j,k+\frac{1}{2}\right)}}\right)H_y^{n-1/2}\left(i+1/2,j,k+\frac{1}{2}\right)$$

$$-\frac{\dfrac{\Delta t}{\mu\left(i+1/2,j,k+\frac{1}{2}\right)}}{1+\dfrac{\sigma_m\left(i+1/2,j,k+\frac{1}{2}\right)\Delta t}{2\mu\left(i+1/2,j,k+\frac{1}{2}\right)}}\left[\frac{E_x^n(i+1/2,j,k+1)-E_x^n(i+1/2,j,k)}{\Delta z}-\frac{E_z^n\left(i+1,j,k+\frac{1}{2}\right)-E_z^n\left(i,j,k+\frac{1}{2}\right)}{\Delta x}\right]$$

(24)

$$E_x^{n+1}\left(i+\frac{1}{2},j,k\right)=\left(\frac{1-\dfrac{\sigma_e\left(i+\frac{1}{2},j,k\right)\Delta t}{2\varepsilon\left(i+\frac{1}{2},j,k\right)}}{1+\dfrac{\sigma_e\left(i+\frac{1}{2},j,k\right)\Delta t}{2\varepsilon\left(i+\frac{1}{2},j,k\right)}}\right)E_x^n\left(i+\frac{1}{2},j,k\right)$$

$$+\frac{\dfrac{\Delta t}{\varepsilon\left(i+\frac{1}{2},j,k\right)}}{1+\dfrac{\sigma_e\left(i+\frac{1}{2},j,k\right)\Delta t}{2\varepsilon\left(i+\frac{1}{2},j,k\right)}}\left[\frac{H_z^{n+\frac{1}{2}}\left(i+\frac{1}{2},j+\frac{1}{2},k\right)-H_z^{n+\frac{1}{2}}\left(i+\frac{1}{2},j-\frac{1}{2},k\right)}{\Delta y}-\frac{H_y^{n+\frac{1}{2}}\left(i+\frac{1}{2},j,k+1/2\right)-H_y^{n+\frac{1}{2}}\left(i+\frac{1}{2},j,k-1/2\right)}{\Delta z}\right]$$

(25)

$$E_y^{n+1}\left(i,j+\frac{1}{2},k\right)=\left(\frac{1-\dfrac{\sigma_e\left(i,j+\frac{1}{2},k\right)\Delta t}{2\varepsilon\left(i,j+\frac{1}{2},k\right)}}{1+\dfrac{\sigma_e\left(i,j+\frac{1}{2},k\right)\Delta t}{2\varepsilon\left(i,j+\frac{1}{2},k\right)}}\right)E_y^n\left(i,j+\frac{1}{2},k\right)$$

$$+\frac{\dfrac{\Delta t}{\varepsilon\left(i,j+\frac{1}{2},k\right)}}{1+\dfrac{\sigma_e\left(i,j+\frac{1}{2},k\right)\Delta t}{2\varepsilon\left(i,j+\frac{1}{2},k\right)}}\left[\frac{H_x^{n+\frac{1}{2}}\left(i,j+\frac{1}{2},k+1/2\right)-H_x^{n+\frac{1}{2}}\left(i,j+\frac{1}{2},k-1/2\right)}{\Delta z}-\frac{H_z^{n+\frac{1}{2}}\left(i+1/2,j+\frac{1}{2},k\right)-H_z^{n+\frac{1}{2}}\left(i-1/2,j+\frac{1}{2},k\right)}{\Delta x}\right]$$

(26)

$$E_z^{n+1}\left(i,j,k+\frac{1}{2}\right)=E_z^{n}\left(i,j,k+\frac{1}{2}\right)\left(\frac{1-\dfrac{\sigma_e\left(i,j,k+\frac{1}{2}\right)\Delta t}{2\varepsilon\left(i,j,k+\frac{1}{2}\right)}}{1+\dfrac{\sigma_e\left(i,j,k+\frac{1}{2}\right)\Delta t}{2\varepsilon\left(i,j,k+\frac{1}{2}\right)}}\right) \tag{27}$$

$$+\frac{\dfrac{\Delta t}{\varepsilon\left(i,j,k+\frac{1}{2}\right)}}{1+\dfrac{\sigma_e\left(i,j,k+\frac{1}{2}\right)\Delta t}{2\varepsilon\left(i,j,k+\frac{1}{2}\right)}}\left[\begin{array}{l}\dfrac{H_y^{n+\frac{1}{2}}\left(i+\frac{1}{2},j,k+1/2\right)-H_y^{n+\frac{1}{2}}\left(i-\frac{1}{2},j,k+1/2\right)}{\Delta x}\\ -\dfrac{H_x^{n+\frac{1}{2}}\left(i,j+1/2,k+1/2\right)-H_x^{n+\frac{1}{2}}\left(i,j-1/2,k+1/2\right)}{\Delta y}\end{array}\right]$$

Note that (ε, σ_e) and (μ, σ_m) are sampled at the points same as the electric and magnetic fields, respectively. From Eqs. (22)–(27), we see that the values of E-field are updated at time step (n + 1), whereas the values of H-field are updated at (n + ½). Also, three arrows are pointing towards each field; one comes from the previous time step and the other two come from a half-step to the right or left. These equations are alternately considered to find the whole solution iteratively. The FDTD method is referred to as the leapfrog method (Figure 3).

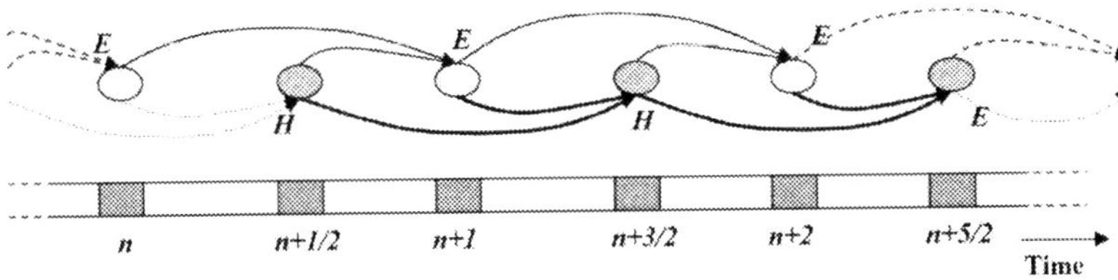

Figure 3. Leapfrog scheme.

4. Absorbing Boundary Conditions

What is known the finite difference equations are solved for a finite computational domain. However, in numerical modeling, many geometries of interest are defined in the "open" regions where the spatial domain of the computed field is unbounded in one or more coordinate directions. Since the data storage in a computer is limited by the size of memory, it is not possible to handle an open region problem directly because more computational time is required. To allay this problem, an absorbing boundary condition (ABC) is used to truncate the computational domain that is designed to suppress spurious reflections of the outgoing waves to an acceptable level.

Although there are various absorbing boundary layer implementations, such as the one derived by Engquist and Majda [8] with the discretization given by Mur [9]. In this group, the perfectly matched layer (PML) technique [5, 10], which was put forward by Berenger in 1994, exhibits an accuracy level that is significantly better than most other ABCs. Consequently, it is widely used in FDTD simulations and is sufficient. Numerical implementation of the PML layer in a 3D FDTD domain is also a straightforward generalization of the 2D case as shown in Figure 4.In the inner vacuum, the finite-difference equations are the usual discretizations of the Maxwell equations. In the PML layer, there are 12 subcomponents to be computed in place of the six components.

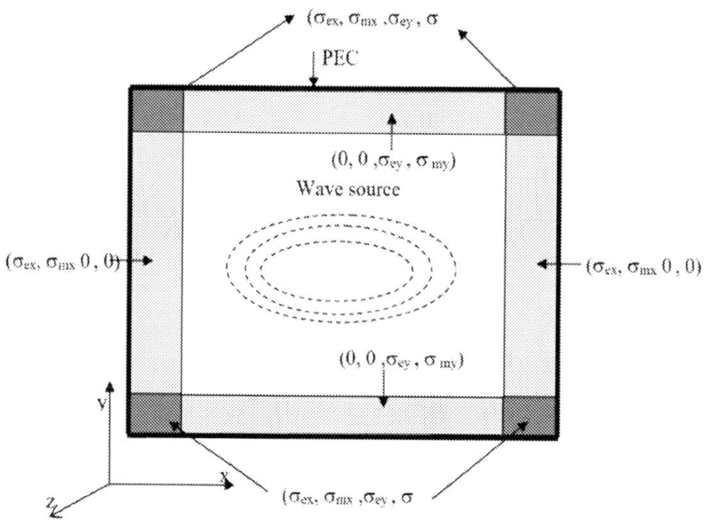

Figure 4. Computational domain surrounded by the PML layer.

In 1994, Berenger [5] proposed non-physical layers of cells defined at the edges of the computational domain, as shown in Figure 4 having special electrical (σ_{ex}, σ_{ey}) and magnetic (σ_{mx}, σ_{my}) conductivities to absorb incident waves from all angles and for any frequency. To ensure that there is no reflection from the inner boundary, the conductivities should satisfy the impedance matching condition:

$$\frac{\sigma_m}{\mu} = \frac{\sigma_e}{\varepsilon} \tag{28}$$

where μ and ε are the magnetic permeability and the electric permittivity, respectively. Also, it is seen that, for the interface normal to the x-axis, $\sigma_{ey} = \sigma_{my} = 0$, and $\sigma_{ex} = \sigma_{mx} = 0$ to that normal at the y-axis. Thus, the PML medium can be characterized by the set of parameters (σ_{ex}, σ_{mx}, σ_{ey}, σ_{my}). Also, the PML layers are backed by a thin perfect electric conductor (PEC), as shown in Figure 4. This PEC layer acts as a perfect mirror for the incident waves. In practical computations, the PML has to be a few cells thick with an electrical conductivity increasing from zero at the interface to a value $\sigma_{e,Max}$ at the outer side of the layer (Figure 5).

For optimal absorbing performance, the loss coefficients usually have a gradual increasing profile $\sigma_e(x)$ with distance inside the PML, and a commonly used one is [5]:

$$\sigma_e = \sigma_{e,Max} \cdot \left(\frac{\rho}{\delta}\right)^2 \tag{29}$$

where ρ is a newly defined coordinate, as shown in Figure 4 and δ is the number of PMLs also:

$$\sigma_{e,Max} = -Ln\big(R(0)\big).\sqrt{\frac{\varepsilon_0}{\mu_0}}\,\frac{3}{2.\delta} \tag{30}$$

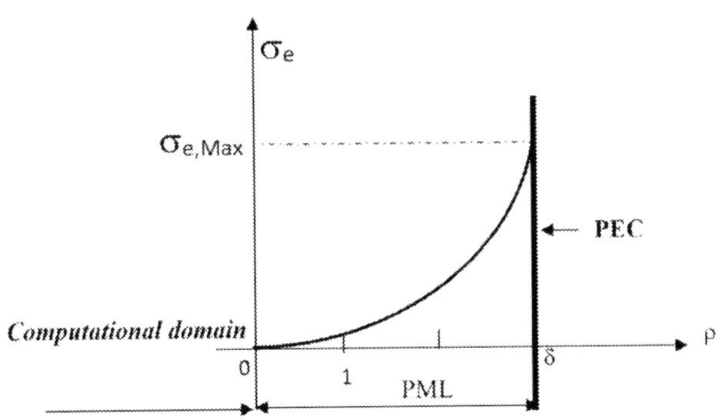

Figure 5. Conductivity profile at the right interface.

The apparent reflection at the normal incidence is $R(0) = e^{-16}$ for $\delta = 10\Delta x$ [5,11,12]. In order to discretize split field Maxwell's equations, we will use the exponential differentiation [13]:

$$a\frac{dA^{n}}{dt}+bA^{n} \square\, b\left[\frac{A^{n+\frac{1}{2}}-A^{n-\frac{1}{2}}e^{\frac{-b\Delta t}{a}}}{1-e^{\frac{-b\Delta t}{a}}}\right] \tag{31}$$

According to Berenger techniques, in the PML medium, each component of the electromagnetic field is split into two parts. In the Cartesian coordinates, the six components yield 12 subcomponents denoted as E_{xy}, E_{xz}, E_{yz}, E_{yx}, E_{zx}, E_{zy}, H_{xy}, H_{xz}, H_{yz}, H_{yx}, H_{zx}, and H_{zy}. Then, the Maxwell equations are replaced by 12 equations. As an example, for the H_z field, we can discretize Eq. (9) in the PML layers as follows:

$$\frac{\partial H_{zx}}{\partial t}+\frac{\sigma_{mx}}{\mu}H_{zx}=-\frac{1}{\mu}\frac{\partial(E_{yx}+E_{yz})}{\partial x} \tag{32}$$

$$\frac{\partial H_{zy}}{\partial t}+\frac{\sigma_{my}}{\mu}H_{zy}=\frac{1}{\mu}\frac{\partial(E_{xy}+E_{xz})}{\partial y} \tag{33}$$

$$\begin{aligned}H_{zx}^{n+1/2}(i+1/2,j+1/2,k)&=H_{zx}^{n-1/2}(i+1/2,j+1/2,k)e^{\frac{-\sigma_{mx}(i+1/2,j+1/2,k).\Delta t}{\mu(i+1/2,j+1/2,k)}}\\&-\frac{1-e^{\frac{-\sigma_{mx}(i+1/2,j+1/2,k).\Delta t}{\mu(i+1/2,j+1/2,k)}}}{\sigma_{mx}(i+1/2,j+1/2,k).\Delta x}.\left[E_{y}^{n}(i+1,j+1/2,k)-E_{y}^{n}(i,j+1/2,k)\right]\end{aligned} \tag{34}$$

$$\begin{aligned}H_{zy}^{n+1/2}(i+1/2,j+1/2,k)&=e^{\frac{-\sigma_{my}(i+1/2,j+1/2,k).\Delta t}{\mu(i+1/2,j+1/2,k)}}H_{zy}^{n-1/2}(i+1/2,j+1/2,k)\\&+\frac{1-e^{\frac{-\sigma_{my}(i+1/2,j+1/2,k).\Delta t}{\mu(i+1/2,j+1/2,k)}}}{\sigma_{my}(i+1/2,j+1/2,k).\Delta y}\left[E_{x}(i+1/2,j+1,k)-E_{x}(i+1/2,j,k)\right]\end{aligned} \tag{35}$$

For more details, readers are referred to [5].

5. Numerical Dispersion and Stability of the FDTDMethod

We cannot choose Δx, Δy, Δz, and Δt arbitrarily without stability problems. For example, investigating Eq. (22) for $\sigma_m = 0$ one may imagine that if the factor $\frac{\Delta t}{\mu \Delta x}$ or $\frac{\Delta t}{\mu \Delta y}$ would be too large, H_z may receive a too large part in each step-in time. H_z could then grow towards infinity.

Moreover, in FDTD, Maxwell's equations are discretized using the second-order accurate centered finite-difference approximation. Therefore, the resultant equations are approximations of the real ones. The obtained difference equations of FDTD converges to the Maxwell's equations when $\Delta x \rightarrow \delta x$, $\Delta y \rightarrow \delta y$, $\Delta z \rightarrow \delta z$ and $\Delta t \rightarrow \delta t$. There are numerical dispersion errors introduced as a result of discretization. Due to this error, the velocity of the wave can be different from the actual velocity, and vary according to its direction of propagation with respect to the grid axis. As an example, the waves traveling along the grid diagonals are slower than the waves propagating along the x and y-axes.

As a result, the precision and algorithmic stability of the FDTD method is governed by the value of Δx, Δy, Δz, and Δt. To minimize the dispersive effects, Eq. (36) should be satisfied [14].

$$Max(\Delta x, \Delta y, \Delta z) \leq \frac{\lambda_{\min}}{10 n_{\max}} \tag{36}$$

where λ_{min} is the minimum wavelength corresponding to the maximum frequency f_{max} of simulation and n_{max} is the maximum refractive index value in the computational domain. For stability, the time steps must be small enough so that the field values can affect only the nearest neighbors during a one-time step interval. The time discretization interval, Δt is determined by the Current Frederic Lewi condition (CFL) [14, 15] given by:

$$c\Delta t \leq \frac{1}{\sqrt{(\frac{1}{\Delta x})^2 + (\frac{1}{\Delta y})^2 + (\frac{1}{\Delta z})^2}} \tag{37}$$

where c is the speed of light in vacuum. When the stability condition is not satisfied, numerical errors grow rapidly as the simulation time evolves, and causes unphysical results. In this chapter, cubical Yee-cells are used, so that $\Delta x = \Delta y = \Delta z = \Delta$ and the temporal time step is given by:

$$\Delta t \leq \frac{\Delta}{c\sqrt{N}} \tag{38}$$

where $N = 1,2,3$ for the 1D, 2D and 3D grids, respectively. In major research works, a conservative choice of Δt is:

$$\Delta t = \frac{\Delta}{2c} \tag{39}$$

6. Source Excitation

To have some use at all for the equations describing time development of the fields, we need either initial values for the fields (i.e., $E(t_0)$ and $H(t_0)$) or some kind of source that adds energy to the system. Then, setting an FDTD computational system with an initial excitation is very important in the implementation of the FDTD algorithm. On the other hand, an improper excitation could lead to spurious solutions that may not physically exist. In addition, the frequency bandwidth of interest of the system can be chosen by simply setting up the width of the time-domain excitation pulse. Usually, a wider time-domain pulse provides a narrow frequency band and vice versa. To avoid undesired noise appearing in the FDTD generated signal responses, it is necessary to choose the excitation pulse smooth and low side lobes. Generally, two types of pulses are used for excitation, the Blackman-Harris and the Gaussian pulses [16]. They are both characterized by smooth transition and side lobes.

In this chapter, a time-dependent point source has been used. The source used in the following simulations is a Gaussian pulse centered at time $t_0 = n_0.\Delta t$.

$$E \;\; or \;\; H \;\; field \;\; \leftarrow Source = e^{-\left[\frac{t-t_0}{\beta.\Delta t}\right]^2} \tag{40}$$

where $\beta \geq 1.273\frac{1}{\Delta t.f_{max}}$ is the decay or pulse width of the source, it will dominate the frequency bandwidth of interest of the system and $n_0 \approx 4\beta$ [17]. Generally, defining a source is as simple as forcing the desired value to points in the grid considered as a source. The source can be a point or a line source or it can have any arbitrary shape.

7. Reflection and Transmission Coefficients

To compute the reflection and transmission coefficients in the frequency domain, the time domain results of the incident, reflected, and transmitted electric or magnetic fields are needed. For each time step, the reflected and transmitted fields are given by:

$$U_{ref}(P_R) = U_{tot}(P_R) - U_{inc}(P_R) \tag{41}$$

$$U_{tran}(P_T) = U_{tot}(P_T) \tag{42}$$

where P_R is the location position for observing the reflected, total, and incident fields, and P_T location is conserved to compute the transmitted field. A better choice of the location indexes among the P_R and P_T is to put P_R at a position between the falling point of the original excitation and the left-most interface of the first dielectric layer to avoid the initial transient effect and edge effect of the first dielectric layer. Another location P_T (the point for computing transmission coefficient) is chosen between the right-most point of the last dielectric layer and the right absorbing boundary.

For the simulation, the electric or magnetic field amplitude at the P_R or P_T locations just in front ($U_{tot}(P_R)$) and at the end ($U_{tran}(P_T)$) of the structure are stored versus time, respectively. We then remove the structure under investigation, and excite the problem again with the free-space everywhere, again saving the electric or magnetic field versus time at the same location as before. This gives us the incident field at the front of the structure ($U_{inc}(P_R)$) versus time. The frequency-domain of the electric or magnetic field can be obtained from the computed time-domain results by using the fast Fourier transform (FFT) of these pulses. Finally, at any specified frequency or frequency band, the reflection coefficient R and the transmission coefficient T can be obtained as follows:

$$R = \frac{FFT(U_{ref})}{FFT(U_{inc})} \tag{43}$$

$$T = \frac{FFT(U_{trans})}{FFT(U_{inc})} \tag{44}$$

Finally, the flowchart of Figure 6 depicts the multistep procedure adopted to use the FDTD method.

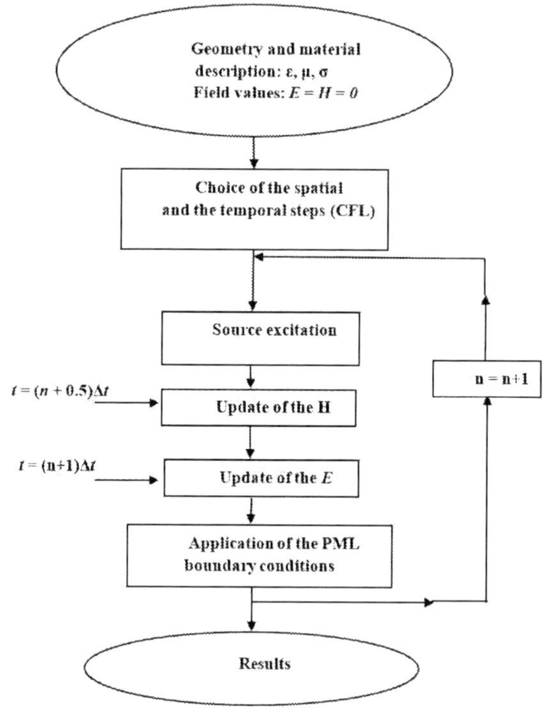

Figure 6. Flowchart of FDTD method.

8. Applications and Related Simulation Results

8.1. Analysis and Design of One-Dimensional Structures

We assume the plane of incidence to be parallel to the *x-o-y* plane. Thus, all field vectors will be dependent on *x* and *y* only, and independent of *z*, since $\partial/\partial z = 0$. The fields in each layer are decomposed into the transverse electric (TE)- and transverse magnetic (TM)-polarization plane waves. For the TE wave, we believe that the electric field is perpendicular to the *x-o-y* plane, and it travels along the x-direction. Thus, the magnetic field oscillates only in the x-y plane. For the TM mode, the electric field oscillates in the x-o-y plane. Therefore, the magnetic field is perpendicular to the x-o-y plane.

8.1.1. FSS Structures Using Inhomogeneous and Stratified Dielectric Multilayer

The first one-dimensional FSS structure considered in this work consists only of five layers.

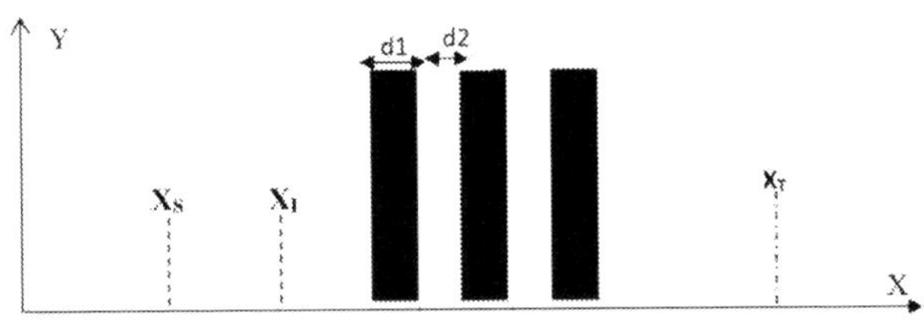

Figure 7. FSS structure.

For odd layers (1, 3, and 5 in black) having the same thickness $d_1 = d_3 = d_5 = 2.3$ mm, the permittivity profile is parabolic and described by

$$\varepsilon_r(x) = \varepsilon_{r\min} + \left(\varepsilon_{r\max} - \varepsilon_{r\min}\right)\sin\left(\frac{\pi.x}{d_{odd}}\right) \tag{45}$$

where $\varepsilon_{rmin} = 9$ and $\varepsilon_{r\,max} = 11$. Even layers having the same thickness $d_2 = d_4 = 7.5$ mm, are assumed to be the air ($\varepsilon_r = 1$). In this work, the FDTD-1D is used with a PML absorbing boundary condition where the code was written using Matlab. The spatial and temporal steps used in the simulation are, respectively, $\Delta x = 1.0050 \times 10^{-4}$ m and $\Delta x = \Delta t/2c$, where c is the wave phase velocity in vacuum.

The Gaussian pulse used for excitation is cantered at time step $n_0 = 280$ and has a decay of $\beta = 70$. The choice of n_0 and β is done considering the range of frequencies covered by the pulse in the region of interest. The time dependence of the source excitation and its Fourier transform is illustrated in Figures 8 and 9, respectively.

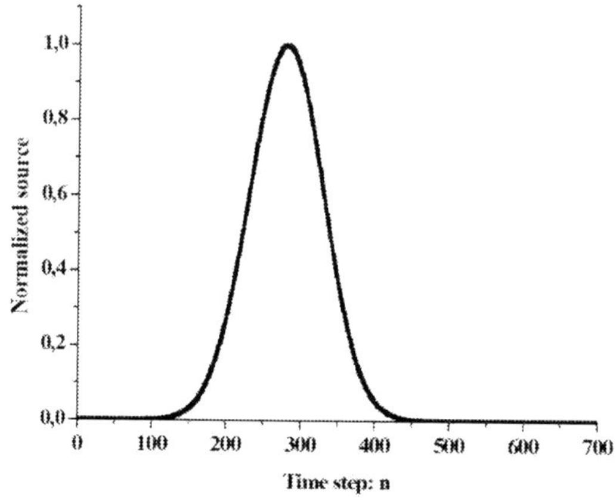

Figure 8. Time dependence of the source.

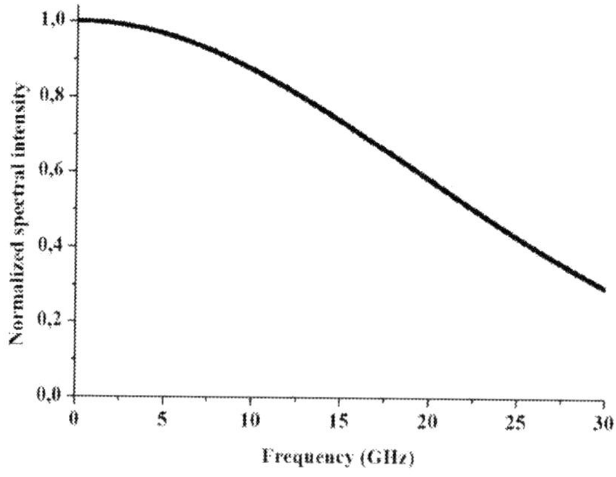

Figure 9. Fourier transform of the source.

The source is launched at the input of the structure at the x-cell $X_S = 140\Delta x$ to avoid its location at the PML interface, as shown in Figure 7. The source cannot remain active during the pulse propagation through the structure, as reflections caused by the layers propagate back to the source and interfere with the launching field, modifying the input conditions. For this purpose, the first layer is put far away from the source to avoid retro reflection. The first layer is placed at the x-cell $I = 400$. The incident, reflected, and transmitted pulses were monitored at the observation points $P_R = X_I = 300$ and $P_T = X_T = 719$ (Figure 7) in front and behind the structure. The time dependence and Fourier transform of these fields are given in Figures 10 and 11, respectively.

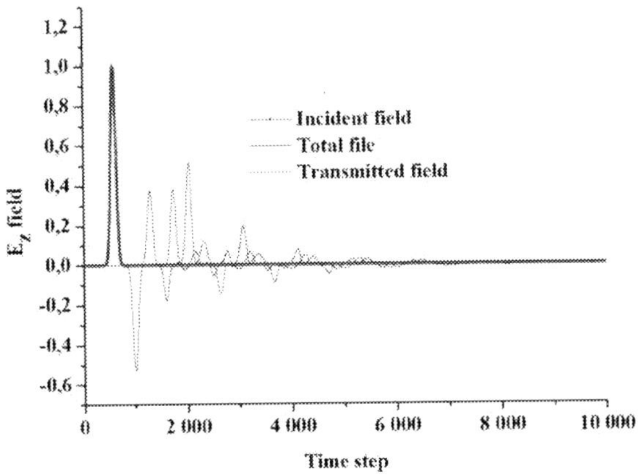

Figure 10. Incident, reflected and transmitted fields.

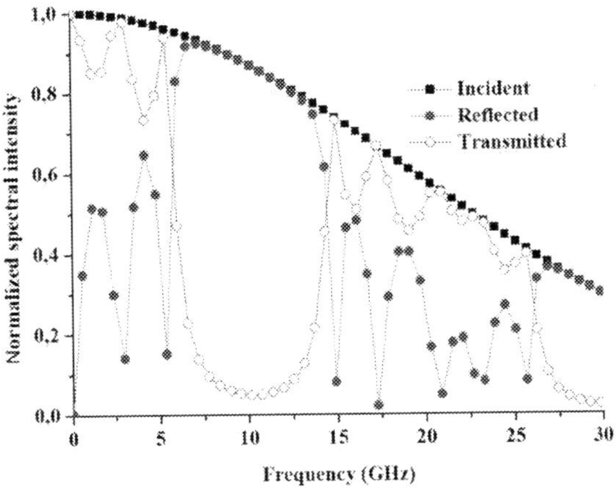

Figure 11. Fourier transform of the incident, reflected, and transmitted fields.

The reflection and transmission coefficients for the TE wave are plotted against frequency and at the normal incidence; Figure 12 illustrates the obtained results. Simulations were done using a PC with a 2.6 GHz clock and 1 Gbyte RAM. Simulated curves show that the bandwidth extending from 6 GHz to 14 GHz is appropriate for high reflection microwave applications. But, the frequencies 3 GHz, 5 GHz, 15 GHz, and 17.5 GHz are suitable for a transparency transmission. Also, ripples in the non-reflective regions are due to the contrast in permittivity between the layers and electrical lengths of lattices.

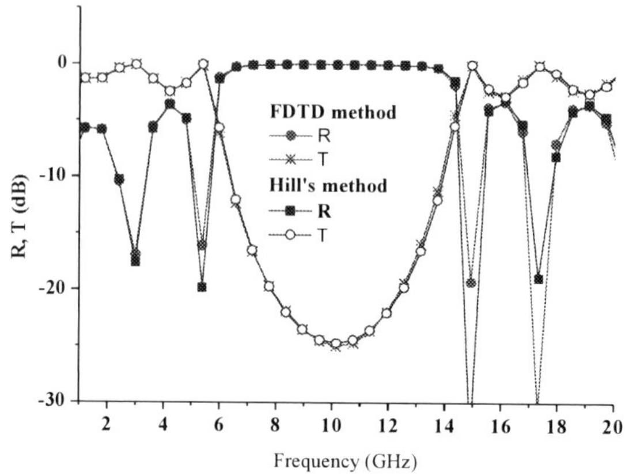

Figure 12. Reflection and transmission coefficients for an FSS.

For a good fitting with other results obtained by Hill's method [6], the time steps used in the calculation by the FDTD method is 10^4. This maximum number of time steps used for the FDTD resolution is not optimized, but it is based on the examination of the resonance peaks on the power spectrum. Afterward, the total calculation time is 7mn, whereas this time is only 51 s when using Hill's method.

The FDTD method has been applied successfully to a multilayered dielectric FSS, and the obtained results agree well with the other obtained by Hill's method [6]. Moreover, according to the above examples, one may conclude that the FDTD method is exact. But, if the simulation time or the number of spatial grids increases, this method is memory intensive and slow because, it stores values for, and operates on, each spatial grid point at each time step.

9. EBG Structures Using Inhomogeneous and Stratified Dielectric Multilayer

The introduction of a defect by breaking the periodicity of the structure under study generates a transmission peak in the stopband region at the frequency f_{Tmax}. This was achieved by widening the central spacing of the central air layer thickness from d_3 to D. Then, an EBG structure is obtained. Figure 13 presents the central frequency f_{Tmax} of the maximum transmission peak versus the defect length D.

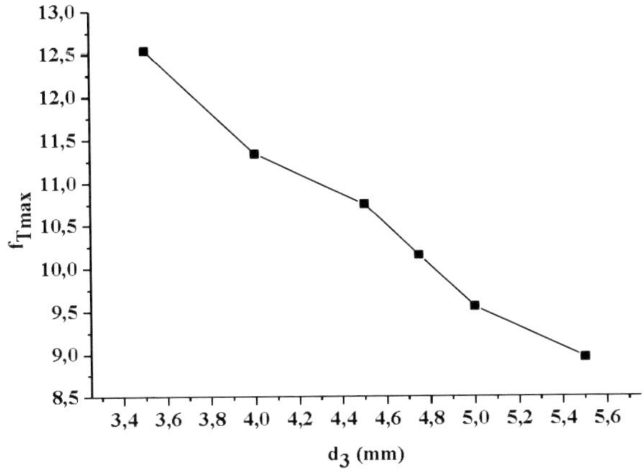

Figure 13. Central frequency f_{Tmax} for each defect length d_3.

For instance, when the defect length is small, the transmission peak shifts to higher frequencies. The frequency shift has its origin in the displacement of the electromagnetic stopband as the structure dimensions are changed. By wisely tuning the defect length, special filter characteristics can be realized. From these results, setting the third air layer thickness $d_3 = 4.8$ mm assuming the frequency of operation as 10 GHz at normal incidence. Figure 14 shows the reflection and transmission characteristics of the designed EBG structure.

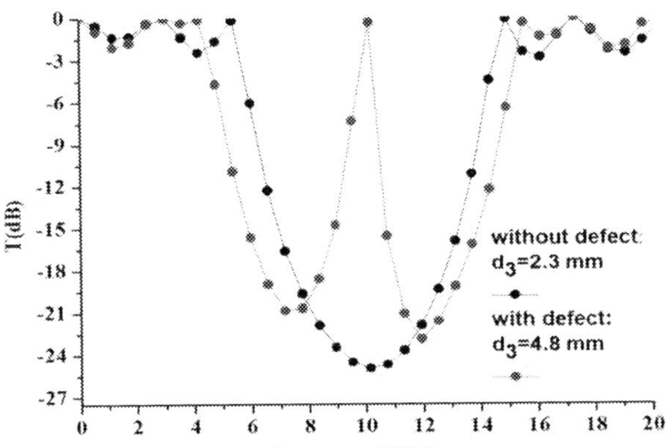

Figure 14. Transmission coefficient for an EBG structure.

Such results show that the structure presents a resonant behavior at the central frequency f = 10 GHz, and can be used as a microwave filter. According to the obtained results in Figure 13, the position of this window can be controlled by choosing the defect length.

10. Analysis and Design Two-Dimensional Structures

10.1. Optical Filter

In this work, the FDTD method is successfully applied to a two-dimensional photonic bandgap (PBG) structure, shown (as an illustrative example) in Figure 15. It is composed of a periodic array of circular holes filled by air in a strip of silicon with a refractive index n = 3.4. To show the effect in the transmission spectrum, changes in geometry (lattice constant A, radius of holes R, and introduction of defect with length D) are introduced.

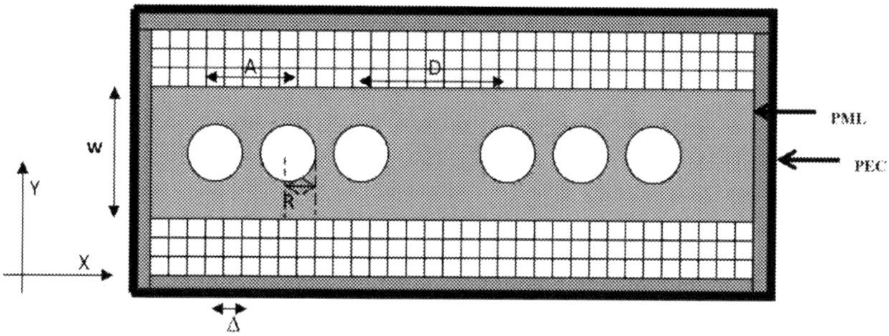

Figure 15. PBG structure and dimensions.

The lattice constant is put equal to A = 364.25 nm; the hole radius R = 131.75 nm and a(D=)503.75nm long defect are introduced at the center, as shown in Figure 15. The structure is assumed as a slab waveguide with a width of w = 434nm and is infinitely large along the z-direction. In FDTD, we have to truncate the mesh in a manner that the exterior boundary looks like an infinite medium. For this aim, we implement PML to absorb waves at the edges of the modelling grid and a Perfect Electric Conductor (PEC) limits the computational domain. In this work, the source excitation is placed at the x-cell is = 20, and it's described by its time and spatial dependence as follows:

$$H_z = H_t(t)H_z(y).$$

The time dependence is given by:

$$H_t(t) = \exp\left\{-\left[(n-n_0)\Delta t\right]^2 / (T\Delta t)^2\right\} \cos(2\pi . f_0 .(n-n_0).\Delta t) \quad (46)$$

where $f_0 = 193.55\times10^{12}$ Hz, $n_0 = 750, T = 250$, $\Delta x = \Delta y = \Delta = \lambda_0/200$, $\Delta t = \frac{\Delta}{c\sqrt{2}}$ and $\lambda_0 = 1550$ nm. The spatial dependence of H_z component is written as follows:

$$H_{z1} = \cos(k.y),\ \text{if } |y| \leq \frac{w}{2} \quad (47)$$

$$H_{z2} = \cos(k.\frac{w}{2})e^{-\gamma(|y|-\frac{w}{2})},\ \text{if } |y| > \frac{w}{2} \quad (48)$$

The propagation parameters for the slab waveguide corresponding to the odds modes are given by [18]:

$$k^2 = [n_1\left(\tfrac{2\pi}{\lambda_0}\right)]^2 - \beta^2 \quad (49)$$

$$\gamma^2 = \beta^2 - [n_0\left(\tfrac{2\pi}{\lambda_0}\right)]^2 \quad (50)$$

The same structure of Figure 15 is used as a filter in optical communication [19]. Its transmittance spectrum is given by Figure 16, where the incident and transmitted pulses are monitored at the observations points $I = 450$ cells and $I = 750$ cells, respectively.

The obtained results show that, when the defect length D is smaller, the holes are closer to each other, and more radiation in the cavity is scattered by them. As a consequence, the intensity of the transmission peak is reduced. As the defect length increases, the holes move apart, and the propagation mode sees less scattering. This way, the intensity of the transmission peak increases. As well, we see a shift in the resonant wavelength to the right, when the defect length increases. In addition, when the radius R of holes decreases (Figures 16. A and 16. C), the resonant wavelength shifts to the right, and the intensity and bandwidth of the transmission peak also increase. Therefore, the designed structure can be used as an optical filter by downscaling the resonant wavelength with controlling the defect length and hole radius.

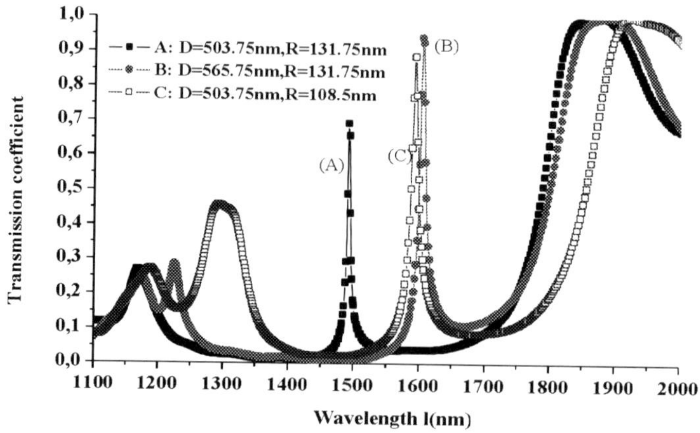

Figure 16. Transmission coefficient and resonant peaks.

10.1.1. Two-dimensional Demultiplexing Structure Based on Photonic Crystals

The structure under investigation is composed of silicon rods ($\varepsilon_r = 12$) in air and arranged in a triangular lattice. The lattice constant is A = 550 nm and the radius rods is $r = 0.15$a. If an entire row of rods is removed, a single line defect waveguide (SLDW) is obtained (Figure 17).

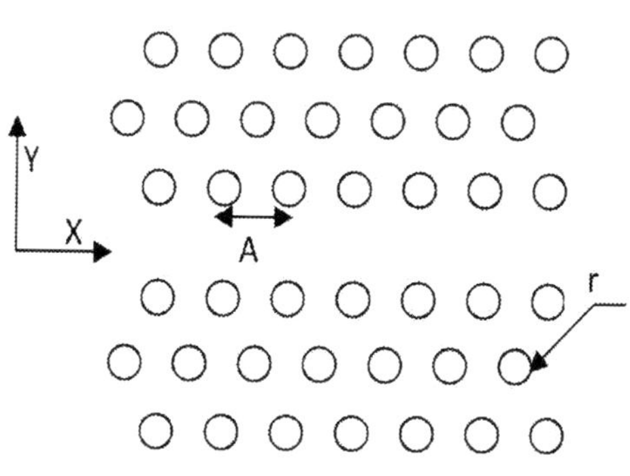

Figure 17. Representation of SLDW.

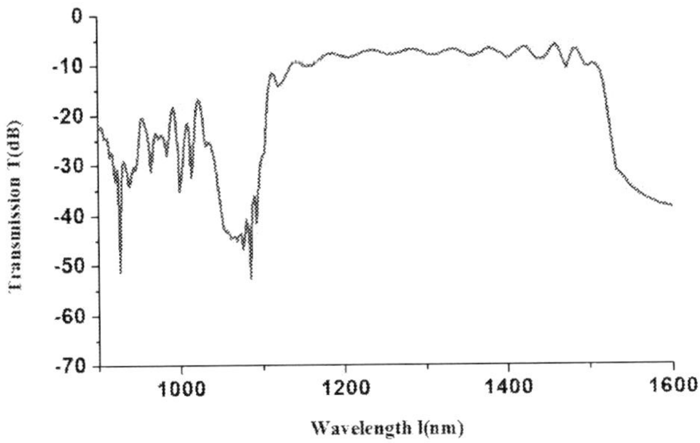

Figure 18. Transmission spectra of the SLDW structure.

A Gaussian wide-band optical pulse is used as the input signal to study the transmission of light through the waveguides. The obtained normalized transmission spectra for the TM wave is shown in Figure 18. It is observed that the SLDW is designed to support the second and third optical transmission windows $\lambda_1 = 1300$ nm and $\lambda_2 = 1500$ nm. If only one of every two rods of the entire row is removed, a coupled-cavity waveguide (CCW) is formed, as shown in Figure 19.

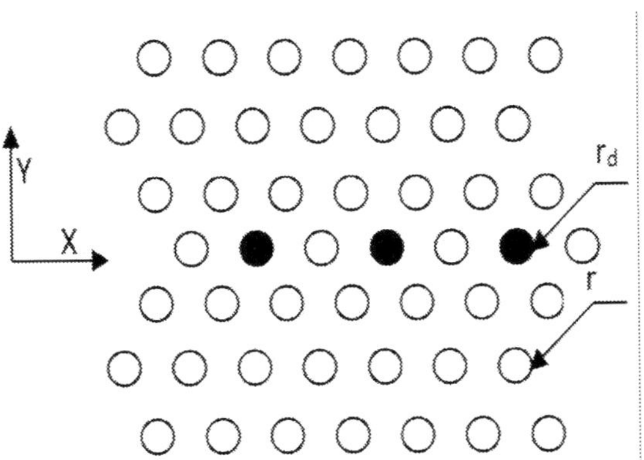

Figure 19. Illustration of CCW structure.

The removed rod is replaced with the other having radius r_d and the same dielectric permittivity (ε_r = 12). Figure 20 shows the plot of the central wavelength λ_{Tmax} of the maximum transmission peak versus the radius r_d of the defect rod. It is seen that $r_d = 0$ and $r_d = 46$ nm will be used to design two

CCWs. So, the CCW_1 with $r_d = 0$ to support a pass-band centered at $\lambda_1 = 1300$ nm, while a CCW_2 with $r_d = 46$ nm to transmit $\lambda_2 = 1550$ nm. Then, the SLDW is coupled to both the CCWs in a Y-form, as shown in Figure 21.

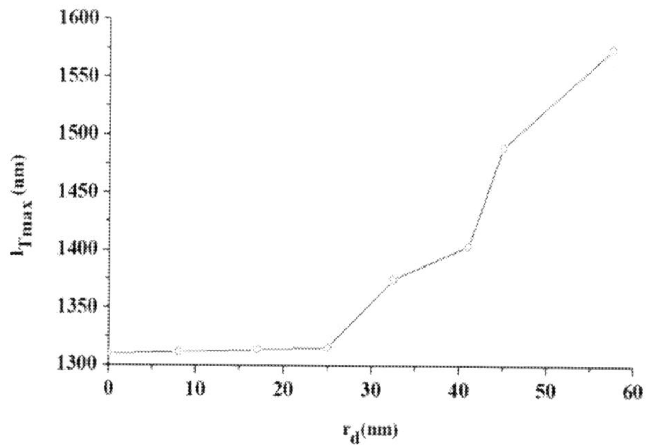

Figure 20. A plot of λ_{Tmax} versus radius r_d of the defect rod.

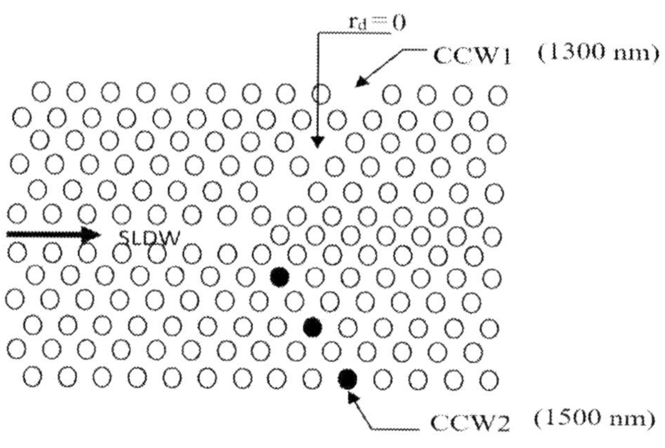

Figure 21. Demultiplexing structure.

So, light travels along an SLDW that afterward is split into two different coupled-cavity waveguides CCW_1 and CCW_2. Then, the SLDW is used to support both the wavelengths, whereas CCW_1 and CCW_2 have been designed to split the second and third optical transmission windows centered at $\lambda_1 = 1300$ nm and $\lambda_2 = 1500$ nm. Figure 22 shows the transmission characteristics for the designed 1×2 demultiplexing structure.

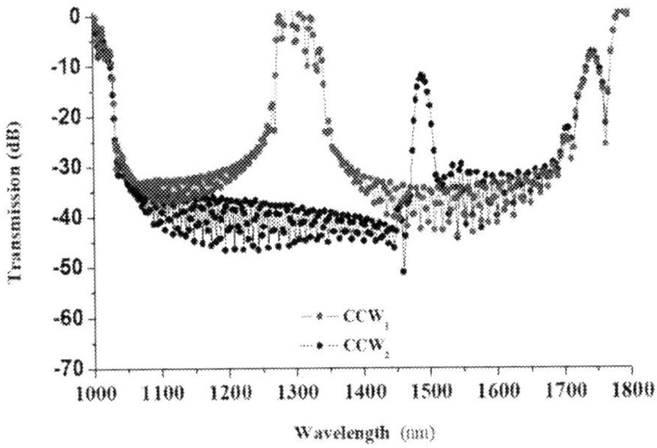

Figure 22. Transmission spectra of the demultiplexing structure.

It is seen that the demultiplexing effect is observed, where the wavelength λ_1 is guided by CCW_1 and λ_2 by CCW_2. But, the losses in both waveguides are observed due to the index mismatching between the input and output ends and the waveguides. Also, the ripples presented in both the transmission spectra are due to the finiteness of waveguides, that is, the finite number of cavities.

This demultiplexing structure may be improved if the ripple due to a finite quantity of cavities is reduced by lengthening the waveguides and the problem of the mismatching between guides will be treated by the matching impedance technique [20]. Also, the same demultiplexing principle described above may be applied to a 2D photonic crystal structure consisting of cylindrical holes on a dielectric substrate.

10.2. Analysis and Design of Two-Channel Wavelength Division Demultiplexer

We build a three-dimensional photonic crystal structure with a square lattice. The background medium was considered as air (i.e., $\varepsilon_r = 1$) and the matrix had cells comprised of silicon cylinders with the relative dielectric permittivity $\varepsilon_r = 9$ (Figure 23). Herein, the square lattice structure under consideration consists of 11 columns × 11 rows of dielectric cylinders. The lattice constant defined as the distance between the two cell centers is $A = 563$ nm and each element radius is $r = 0.2\ A$. The z-direction is taken along the dielectric columns. Along this direction the waveguide dimension is A.

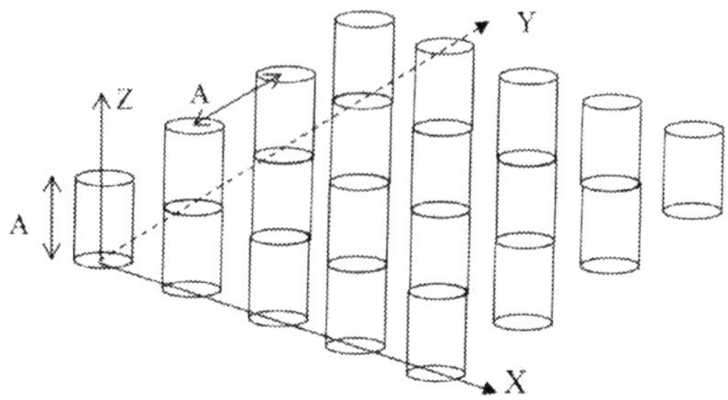

Figure 23. 3D photonic crystal with a square.

Many optoelectronic devices are based on waveguides since they can transport electromagnetic energy. Moreover, the waveguides coupled with resonators may lead to other new devices based on a series of photonic systems. For the next device, the defect in crystal lattice was created by removing one column of rods along the *X*- and *Y*-directions, afterward, one can obtain a T bent waveguide like the one in Figure 24(a). Wavelength demultiplexing is carried out using a two-channel splitter F_1 and F_2 created by three successive rods with radius $r_1 = 0.12A$ and $r_2 = 0.28A$, respectively. The filter F_1 with $r_1 < r$ fully reflects the radiation with 1310 nm wavelength and F_2 with $r_2 > r$ – the radiation with 1550 nm wavelength. Moreover, to achieve excellent guiding performance [20], one column is added and another is removed at the corners of each T branch (Figure 24(b)).

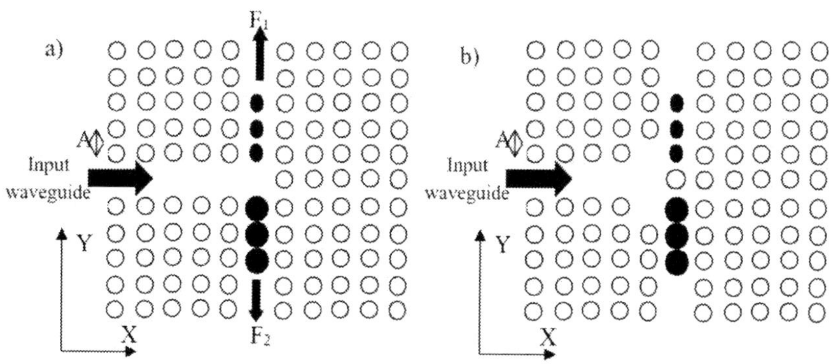

Figure 24. 2D view of the T-shaped demultiplexer.

In FDTD simulations with a PML as absorbing boundary conditions, the spatial grid size is $\Delta x = \Delta y = \Delta z = A/21$ and the temporal step Δt is calculated using the CFL condition (Eq. (38)). The Gaussian pulse used for excitation is centered at the time step $n_0 = 4\beta$, and has a decay of $\beta = 20$. After 10000 steps in time, the transmission spectra versus wavelength is shown in Figure 25.

It is seen that for both cases, the transmittance for the channel 1550 nm is better than for the 1310 nm filter. Figure 21 shows the comparison of the initial T-structure to the finetuned one by transmission spectra. As one can see, the tuning at T junctions provides better spectra and excellent guiding performance, which makes it possible to decrease the reflectivity of the whole structure at the operating wavelengths. The issue is that one may wish to move the optical power from one position in space to another; Figure 25 reveals that this is possible.

To understand better the splitting phenomenon and the advantage of the FDTD method, the computed electric field $E_z(i, j, k = 10)$ distribution in the square lattice structure at time $t = 2000\Delta t$ is shown in Figure 26 for the transmitted signal with 1310 nm and 1550 nm wavelengths.

The wavelength $\lambda_1 = 1550$ nm propagates only in filter F_1 with $r_1 < r$, and the wavelength $\lambda_2 = 1310$ nm transfers to filter F_2 through the coupling region. Also, the field distribution is nonuniform for both the wavelengths, and parasitic penetration of the radiation through the filters into the adjacent secondary waveguide is observed. To overcome this problem is not trivial, since various interactions, such as coupling, reflection, interference, or diffraction, are present at the coupling region.

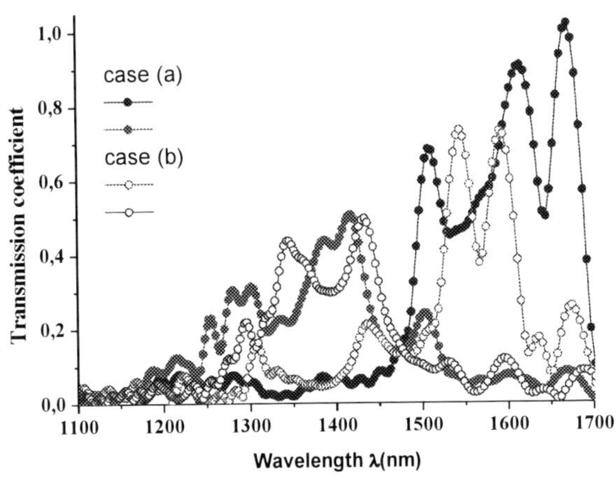

Figure 25. Transmission spectra of the T-shaped demultiplexer corresponding to the structures in Figure 24 (a & b).

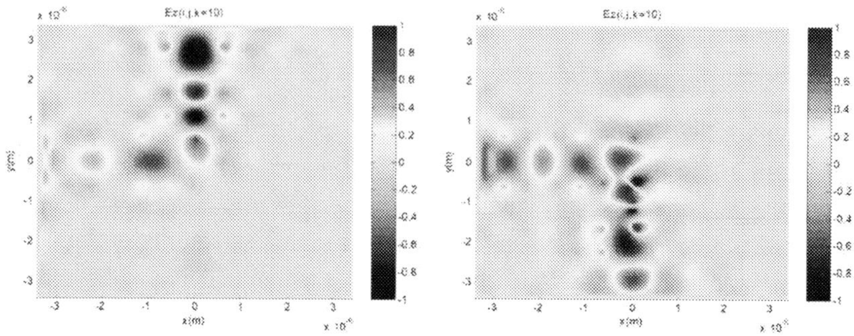

Figure 26. The distribution shape of the electric field E_z $(i, j, 10)$ at time $t = 2000\Delta t$.

It has been shown previously that even more complicated 2D and 3D photonic crystal configurations can be modeled with the FDTD approach. More complex materials and structures are readily included. This flexibility of the FDTD simulator makes it a very useful design tool for use in the investigation of photonic crystals devices, whether they are studied in the microwave, millimeter-wave, or optical wave regimes.

Conclusion

This chapter briefly reviewed the fundamentals of the FDTD method, including Yee's spatial and temporal grid and the updating formulations. This is done along with the implementation of an artificial absorbing boundary medium known as the PML. This medium enables to limit of the computational domain required for simulation. In this work, numerical examples are simulated and compared with those using Hill's method. Furthermore, several applications have been investigated, such as the problem of modeling the FSS and EBG structures, optical filters, waveguides, and demultiplexer based on photonic crystals. These devices are modeled in terms of the transmission and reflection characteristics for the TE and TM waves.

References

[1] Andonegui and A. J. Garcia-Adeva, "The finite element method applied to the study of two-dimensional photonic crystals and resonant cavities," *Opt. Express*, vol. 21, no. 4, pp. 4072–4092, 2013.

[2] Vardaxoglou, J. C. "*Frequency Selective Surfaces,*" Research Studies Press: Boston, USA, (1997).

[3] Yee, K. S. "Numerical solution of initial boundary value problems involving Maxwell's equations in isotropic media," *IEEE Trans. Antennas and Propag.*, vol. 14, no. 3, pp. 302-307, 1966.

[4] Taflove, A. "*Computational Electrodynamics: The Finite-Difference Time-Domain Method,*" Artech House INC, Norwood, 1995.

[5] Berenger, J. P. "A perfectly matched layer for the adsorption of electromagnetic waves," *J. Comput. Phys.*, vol. 114, no. 2, pp. 185-200, 1994.

[6] Choubani, M., F. Choubani, A. Gharsallah, J. David, N. E. Mastorakis, "Analysis and design of electromagnetic band gap structures with stratified and inhomogeneous media," *Opt. Commun.* Vol. 283, no. 22, pp. 4499-4504, 2010.

[7] Sadiku, M. N. O. "*Numerical Techniques in Electromagnetic*" 3rd Ed., CRC- Press, Boka Raton (2001).

[8] Engquist B. and A. Majda, ''Absorbing boundary conditions for the numerical simulations of waves,'' *Proc. Natl. Acad. Sci. USA.*, vol. 74, no. 5, pp. 1765-1766, 1977:

[9] Mur, G. "Absorbing boundary conditions for the finite difference approximation of the time domain electromagnetic-field equations," *IEEE Trans. Electromagn. Compat.*, vol. 4, pp. 377-382, 1981.

[10] Berenger, J. P. "Three-Dimensional Perfect Matched Layer for the Absorption of Electromagnetic Waves," *J. Comput. Phys.*, vol. 127, no. 2, pp. 363-379, 1996.

[11] Gedney, S. D. "An anisotropic perfectly matched layer absorbing media for the truncation of FDTD lattices" *IEEE Trans. Antennas and Propag.*, Vol. 44, no. 12, pp. 1630-1639, 1996.

[12] He., L. "*FDTD – Advances in Sub-Sampling Methods, UPML, and Higherorder Boundary Conditions,*" M. S. thesis, University of Kentucky, 1997.

[13] Yuan, X., D. Borup, J. W. Wiskin, M. Berggren, R. Eidens, S. A. Johnson, "Formulation and validation of Berenger's PML absorbing boundary for the FDTD simulation of acoustic scattering," *IEEE Trans. Ultrason. Ferroelectr. Freq. Control*, vol. 44, no. 4, pp. 816-822, 1997.

[14] Taflove A. and S. Hagness, "*Computational Electrodynamics: The Finite – Difference Time– Domain method*", 3rd Ed., Artech House, Inc., Norwood, MA (2005).

[15] Taflove A. and Morris E. Brodwin., "Numerical solution of steady- state electromagnetic scattering problems using the time- dependent Maxwell's equations," *IEEE Trans. Microwave Theory Tech.*, vol. 23, no. 8, pp. 623-630, 1975.

[16] Kunz K. S. and R. J. Luebbers, "*The Finite Difference Time Domain Method for Electromagnetic,*" CRC Press, Inc., New York (1993).

[17] Chang S. and Robert. Nevels, "Optimizing the Gaussian excitation function in the Finite Difference Time Domain method," *IEEE Trans. Educ.*, vol. 45, no. 1, pp.15-18, 2002.

[18] Cherin, H. "*An introduction to Optical Fibers,*" McGraw-Hill Book Co., New York (1983).

[19] de Tarso Neves, P. Jr. and Alexandre A. P. Pohl,. "Bandwidth Analysis of Silicon PhC filters or CWDM Systems Using FDTD-2D method," *J. Commun. Inform. Syst.*, vol. 20, no.3, pp. 96-104, 2005.

[20] Boscolo, S., C. Conti, M. Midrio, and C. G. Someda, "Numerical Analysis of Propagation and Impedance Matching in 2-D Photonic Crystal Waveguides With Finite Length," *J. Light. Technol.*, vol. 20, no. 2, pp.304–310, 2002.

Chapter 5

Nonlinear Optical Properties in Single and Coupled InAs/GaAs Quantum Dots

Mohsen Choubani*, Hassen Maaref and Faouzi Saidi

Laboratoire de Micro-Optoélectronique et Nanostructures (LR99ES29), Département de Physique, Faculté des Sciences de Monastir, Université de Monastir, Tunisia

Abstract

In this chapter, we explore the effects of structural dimensions, wetting layer, applied electric field, hydrostatic pressure, and temperature on the linear and nonlinear optical properties (NLOP) in single (SQD), coupled InAs/GaAs quantum dots (CQDs), and core/shell. The analytical expressions of the NLOP were analyzed by using the density matrix formalism, effective mass, and finite difference method (FDM). The obtained results showed that the NLOP obtained with these systems is not a monotonic function of the barrier width, electric field, pressure, and temperature. Also, the results revealed that the resonant peaks of NLOP can be blue-shifted or red-shifted depending on the energy of the lowest confined states in such structures. In addition, this condition can be controlled by changes in structural dimensions and the external proofs mentioned above.

Keywords: quantum core/shell, single and coupled quantum dots, electric field, nonlinear optical properties

---

* Corresponding Author's E-mail: mohsenchoubani3@yahoo.fr.

In: Second Harmonic Generation
Editors: Abdel-Baset M. A. Ibrahim and Pankaj Kumar Choudhury
ISBN: 978-1-68507-888-1

1. Introduction

Semiconductor nanostructures, known as quantum dots, confine electrons on the nanometer length scale in all spatial directions. Therefore, QDs may be considered as artificial atoms with discrete energy levels for electrons. In the past few years, nonlinear optical properties related to intersubband transitions in QDs have attracted more attention on theoretical and experimental sides [1–3]. In many investigations, the QD shapes were approximated as pyramidal [4], dome-shaped [5], and elliptic [6]. Recently, the QD shape has been considered as a convex-plane lens [7]. In this chapter, the lens-shaped QDs is considered.

Besides, during the Stransky-Krastanov growth method, after reaching the active layer to a critical thickness, the strain-driven growing layer turns into a three-dimensional island, leaving a thin layer known as the wetting layer (WL). This is because a lattice mismatching of about 7% in an InAs/GaAs system occurs during the epitaxial growth of InAs on GaAs substrate. In this case, the WL thickness is about 0.56 nm [8].

Recently, much attention has been focused on nonlinear optical properties, such as Second Harmonic Generation (SHG) [9], Third Harmonic Generation (THG) [10], and Nonlinear Optical Rectification (NOR) [11]. Also, different shapes of QDs have been studied over the years. In the literature, a single QD with a WL or a coupled QDs without WL are investigated [12, 13]. However, in this work, a double coupled lens-shaped InAs QDs embedded in a GaAs matrix with wetting layers are also considered under combined effects of external electric and magnetic fields, pressure, and temperature.

This chapter is organized as – in Section 2, the Hamiltonian and confinement potential are discussed and given in the effective mass approximation. The Hamiltonian, relevant eigen-states, and Eigen-energies are discussed and numerically determined by using the finite difference method (FDM). In Section 3, we search to test and validate our implemented code by using an ideal cubic QD system. In Section 4, calculations are carried out for a single quantum dot (SQD) system to further understand and show the efficacy of using the FDM to optimize the NOR coefficient. In Section 5, after obtaining the energies and their corresponding envelope wave functions, we have taken the opportunity to generalize our implemented code for laterally coupled quantum dots (LCQDs) grown on a wetting layer (WL) and aligned in the same plane to achieve NLOP in the terahertz domain (1.2–2.88 THz). Here, the numerical results are obtained under the effects of hydrostatic pressure and temperature. In Section 6, numerical simulations are carried out for typical vertically coupled QDs (VCQDs) along the direction of growth (z-direction) and under an electric field. In Section 7, a lens-shaped quantum core/shell (LSQCS) system coupled to a WL and under transversal applied electric field is investigated. Finally, Section 8 summarizes a conclusion.

2. Theory

2.1. Schrödinger Equation

In the framework of effective mass and envelope function approxima-tions, the three-dimensional time-independent Schrödinger equation for a confined electron or hole in such a structure is given by [14]

$$-\frac{\hbar^2}{2}\left[\nabla\left(\frac{1}{m^*(x,y,z,P,T)}\nabla\right)+V(x,y,z,P,T)+e\vec{F}\,\vec{r}\right]\psi(x,y,y,z)=E\psi(x,y,y,z) \tag{1}$$

where m^* represents the electron effective mass, $\vec{F}$ is the external applied electric field and V is the confining potential. Unlike the published works, we have introduced the position-dependent $\vec{r}(x,y,z)$, pressure (P) and temperature (T) dependence of the electron effective mass given by the Vurgaftman formula [15]:

$$\frac{m_0}{m^*(x,y,z,P,T)}=1+2\gamma+\left[\frac{E_P^{\Gamma}\left[E_g^{\Gamma}(P,T)+\frac{2}{3}\Delta_{SO}\right]}{E_g^{\Gamma}(P,T)\left[E_g^{\Gamma}(P,T)+\Delta_{SO}\right]}\right] \tag{2}$$

where m_0 is the free-electron mass, γ is the Kane parameter, E_P^{Γ} is the energy related to the momentum matrix element, Δ_{SO} is the spin-orbit splitting, and E_g^{Γ} is the energy gap at Γ-point. The energy gaps in InAs and GaAs semiconductor materials are temperature- and pressure-dependent quantities, with the functional form as in Eq. (3) [16]:

$$E_g^{\Gamma}(P,T) = E_g^{\Gamma}(0,0) - \frac{\alpha T^2}{\beta + T} + D.P \tag{3}$$

where $E_g^{\Gamma}(0,0)$ is the energy gap at pressure $P = 0$ Kbar and temperature $T =$ 0K. (α, β) and D are the temperature and pressure coefficients, respectively. The 3D confining potential energy as a function of spatial coordinates, pressure, and temperature is given by

$$V(x,y,z,P,T) = \begin{cases} 0, \text{ inside InAs QDs } \mathit{and} \ \mathit{WL} \\ Q_c \Delta E_g^{\Gamma}, \text{ elsewhere} \end{cases} \tag{4}$$

where Q_C is the conducting band offset parameter and $\Delta E_g^{\Gamma} = E_g^{\Gamma}(GaAs) - E_g^{\Gamma}(InAs)$ is the bandgap difference between GaAs and InAs materials at the Γ-point. The recommended parameters used in Eqs. (1) – (3) and adopted in our calculations are illustrated in Table 1 for GaAs and InAs [15, 16].

Table.1. Recommended parameters for GaAs and InAs

Parameters	GaAs	InAs
$E_g^{\Gamma}(0,0)$ (eV)	1.519	0.533
α (meV/K)	0.5405	0.276
β (K)	204	83
D (meV/Kbar)	10.8	7.7
Δ_{SO} (eV)	0.341	0.39
E_p^{Γ} (eV)	28.8	21.5
γ	–1.94	–2.9

The Schrödinger equation of the considered system cannot be solved analytically. Therefore, the FDM is used in this work to calculate the direct numerical solutions for Eq. (1). To solve this equation numerically over a three-dimensional volume (L_x, L_y, L_z), the structure is discretized into $N_x = N_y = N_z$ nodes along the x-, y- and z-directions, respectively. For simplicity, the distance between two consecutive nodes is $\Delta x = \Delta y = \Delta z = \Delta$. It is then required to find the value of wave function at each of these grid points. Let $\psi(x_i, y_j, z_k)$

= $\psi(i, j, k)$ represents the value of the wave function at the grid point (x_i, y_j, z_k) where i, j, and k = 1, 2, ... N_z.

By using the central finite difference approximation for each of the second-order derivatives, we get:

$$\frac{\partial\psi(i,j,k)}{\partial x}=\frac{\psi(i+1,j,k)-\psi(i-1,j,k)}{2\Delta}$$
$$\frac{\partial\psi(i,j,k)}{\partial y}=\frac{\psi(i,j+1,k)-\psi(i,j-1,k)}{2\Delta} \quad (5)$$
$$\frac{\partial\psi(i,j,k)}{\partial z}=\frac{\psi(i,j,k+1)-\psi(i,j,k-1)}{2\Delta}$$

The finite-difference representation of Eq. (1) is as follows:

$$\begin{aligned}&A(i,j,k)\psi(i,j,k)+B(i,j+1,k)\psi(i,j+1,k)+C(i+1,j,k)\psi(i+1,j,k)\\&+D(i,j-1,k)\psi(i,j-1,k)+E(i-1,j,k)\psi(i-1,j,k)\\&+F(i,j,k+1)\psi(i,j,k+1)+G(i,j,k-1)\psi(i,j,k-1)=E\psi(i,j,k)\end{aligned} \quad (6)$$

However, the standard boundary conditions still need to be applied, which is:

$$\psi(i,j,k)\rightarrow 0 \quad as\, x,y,z\rightarrow\pm\infty \quad (7)$$

This is achieved by forcing the value of wave function to be zero at the faces of a three-dimensional cube of grid points. This effectively means that we only need to find the wave function at the interior points of the 3D grid. For each of the grid points, there exists a linear equation. Consequently, there are $(N_z-2)^3$ linear equations that are converted into the matrix eigenvalue problem, $H\psi = E\psi$. For example, the matrix representation of H for a 4×4×4 points are given as follows:

$$H=\begin{bmatrix}\begin{pmatrix}A&B&0&0\\D&A&B&0\\0&D&A&B\\0&0&D&A\end{pmatrix}&\begin{pmatrix}C&0&0&0\\0&C&0&0\\0&0&C&0\\0&0&0&C\end{pmatrix}&\begin{pmatrix}0&0&0&0\\0&0&0&0\\0&0&0&0\\0&0&0&0\end{pmatrix}&\begin{pmatrix}F&0&0&0\\0&F&0&0\\0&0&F&0\\0&0&0&F\end{pmatrix}\\\begin{pmatrix}E&0&0&0\\0&E&0&0\\0&0&E&0\\0&0&0&E\end{pmatrix}&\begin{pmatrix}A&B&0&0\\D&A&B&0\\0&D&A&B\\0&0&D&A\end{pmatrix}&\begin{pmatrix}C&0&0&0\\0&C&0&0\\0&0&C&0\\0&0&0&C\end{pmatrix}&\begin{pmatrix}0&0&0&0\\0&0&0&0\\0&0&0&0\\0&0&0&0\end{pmatrix}\\\begin{pmatrix}0&0&0&0\\0&0&0&0\\0&0&0&0\\0&0&0&0\end{pmatrix}&\begin{pmatrix}E&0&0&0\\0&E&0&0\\0&0&E&0\\0&0&0&E\end{pmatrix}&\begin{pmatrix}A&B&0&0\\D&A&B&0\\0&D&A&B\\0&0&D&A\end{pmatrix}&\begin{pmatrix}C&0&0&0\\0&C&0&0\\0&0&C&0\\0&0&0&C\end{pmatrix}\\\begin{pmatrix}G&0&0&0\\0&G&0&0\\0&0&G&0\\0&0&0&G\end{pmatrix}&\begin{pmatrix}0&0&0&0\\0&0&0&0\\0&0&0&0\\0&0&0&0\end{pmatrix}&\begin{pmatrix}E&0&0&0\\0&E&0&0\\0&0&E&0\\0&0&0&E\end{pmatrix}&\begin{pmatrix}A&B&0&0\\D&A&B&0\\0&D&A&B\\0&0&D&A\end{pmatrix}\end{bmatrix} \tag{8}$$

where:

$$\begin{aligned}
A(i,j,k)&=\frac{-\hbar^2}{4\Delta^2}\left\{\begin{aligned}&\frac{6}{m^*(i,j,k)}+\frac{1}{m^*(i-1,j,k)}+\frac{1}{m^*(i+1,j,k)}\\&+\frac{1}{m^*(i,j-1,k)}+\frac{1}{m^*(i,j+1,k)}\\&\frac{1}{m^*(i,j,k-1)}+\frac{1}{m^*(i,j,k+1)}\end{aligned}\right\}+V(i,j,k)\\
B(i,j+1,k)&=\frac{-\hbar^2}{4\Delta^2}\left\{\frac{1}{m^*(i,j,k)}+\frac{1}{m^*(i,j+1,k)}\right\}\\
C(i+1,j,k)&=\frac{-\hbar^2}{4\Delta^2}\left\{\frac{1}{m^*(i,j,k)}+\frac{1}{m^*(i+1,j,k)}\right\}\\
D(i,j-1,k)&=\frac{-\hbar^2}{4\Delta^2}\left\{\frac{1}{m^*(i,j,k)}+\frac{1}{m^*(i,j-1,k)}\right\}\\
E(i-1,j,k)&=\frac{-\hbar^2}{4\Delta^2}\left\{\frac{1}{m^*(i,j,k)}+\frac{1}{m^*(i-1,j,k)}\right\}\\
F(i,j,k+1)&=\frac{-\hbar^2}{4\Delta^2}\left\{\frac{1}{m^*(i,j,k)}+\frac{1}{m^*(i,j,k+1)}\right\}\\
G(i,j,k-1)&=\frac{-\hbar^2}{4\Delta^2}\left\{\frac{1}{m^*(i,j,k)}+\frac{1}{m^*(i,j,k-1)}\right\}
\end{aligned} \tag{9}$$

In this work, the sparse matrix is used to implicitly solve the system of equations given by Eq. (8). After the diagonalization of the Hamiltonian H, one can calculate the direct numerical solutions for Eq (1), such as the electron energy states (E_i) and their matching envelope wave functions (Ψ_i). Thus, in this work, the Schrödinger equation is solved implicitly by using the FDM of second order.

2.2. Nonlinear Optical Properties

Under an incident electromagnetic field with frequency ω, an electronic polarization $P(t)$ of the system is generated. The incident field and the macroscopic polarization are defined as follows:

$$E(t) = \tilde{E}_0 e^{i\omega t} + \tilde{E}_0 e^{-i\omega t} \tag{10}$$

$$P(t) = \varepsilon_0 \chi^{(1)}(\omega)\tilde{E}_0 e^{i\omega t} + \varepsilon_0 \chi_{2\omega}^{(2)}(2\omega)\tilde{E}_0^{2} e^{2i\omega t} + \varepsilon_0 \chi_{3\omega}^{(3)}(3\omega)\tilde{E}_0^{3} e^{3i\omega t} + cc + \varepsilon_0 \chi_0^{(2)} \left|\tilde{E}_0\right|^2 \tag{11}$$

Here the terms $\chi^{(1)}, \chi_{2\omega}^{(2)}(2\omega), \chi_{3\omega}^{(3)}(3\omega)$ are the linear, the second, and the third-order nonlinear optical susceptibilities, respectively.

2.2.1. Optical Absorption Coefficients

The calculations of the linear ($\alpha^{(1)}$), the third-order nonlinear ($\alpha^{(3)}(\omega, I)$) and the total (α) optical absorption coefficients are based on the density matrix approach, for which the total absorption coefficient (AC) is given by [17, 18]:

$$\alpha(\omega, I) = \alpha^{(1)}(\omega) + \alpha^{(3)}(\omega, I) = \omega\sqrt{\frac{\mu}{\varepsilon_0 \varepsilon_r}}\,\mathrm{Im}\left[\varepsilon_0 \chi^{(1)}(\omega) + \varepsilon_0 \chi^{(3)}(\omega)\right] \tag{12.a}$$

where:

$$\alpha^{(1)}(\omega) = \omega\sqrt{\frac{\mu}{\varepsilon_0 \varepsilon_r}}\left[N\hbar\Gamma_{if}\left|\mu_{if}\right|^2 \Big/ (E_{fi} - \hbar\omega)^2 + (\hbar\Gamma_{if})^2\right] \tag{12.b}$$

and:

$$\alpha^{(3)}(\omega,I) = -\omega\sqrt{\frac{\mu}{\varepsilon_0\varepsilon_r}}\frac{I}{2\varepsilon_0 n_r c}\left[N\hbar\Gamma_{if}\left|\mu_{if}\right|^4 \Big/ \left[(E_{fi}-\hbar\omega)^2+(\hbar\Gamma_{if})^2\right]^2\right]\times$$
$$\left\{4-\left[(\mu_{ff}-\mu_{ii})^2\left(3E_{fi}^2-4E_{fi}\hbar\omega+(\hbar\omega)^2-(\hbar\Gamma_{if})^2\right)\Big/\left|\mu_{if}\right|^2\left(E_{fi}^2+(\hbar\Gamma_{if})^2\right)\right]\right\} \tag{12c}$$

Here, $\hbar\omega$ is the incident photon energy, $I = 2\varepsilon_0 n_r c\left|\tilde{F}\right|^2$ is the incident optical intensity, $n_r = \sqrt{\varepsilon_r} = 3.2$ is the refractive index, $\mu_{ij} = \langle\psi_i|ez|\psi_j\rangle$ is the dipole matrix element, $N = 3\times10^{22}$ m^{-3} is the volume density of carriers in the system, $E_{fi} = E_f - E_i$ is the transition energy between the excited (E_f) and ground (E_i) energy states, $\Gamma_{if} = \frac{1}{T_{if}} = 2ps^{-1}$ [19, 20] is the relaxation rate between excited and ground states. For more details about the density matrix approach, researchers are referred to [21].

2.2.2. Nonlinear Optical Rectification

After obtaining the energies E and their corresponding envelope wave functions Ψ, the NOR can be calculated as follows [22]:

$$\chi_0^{(2)} = \frac{4e^3N\mu_{12}^2\delta_{21}}{\varepsilon_0}\frac{E_{21}^2\left[1+\frac{\Gamma_2}{\Gamma_1}\right]+\left[(\hbar\omega)^2+(\hbar\Gamma_2)^2\right]\left[\frac{\Gamma_2}{\Gamma_1}-1\right]}{\left[(E_{21}-\hbar\omega)^2+(\hbar\Gamma_2)^2\right]\left[(E_{21}+\hbar\omega)^2+(\hbar\Gamma_2)^2\right]} \tag{13}$$

where $N = 3\times10^{22}$ m^{-3} is the volume density of carriers, $\delta_{21} = |\mu_{22}-\mu_{11}|$, where $\mu_{ij} = \langle\psi_i|z|\psi_j\rangle$ is the dipole matrix element, $E_{21} = (E_2 - E_1)$ is the transition energy between the first excited state and the ground state, $\hbar\omega$ is the incident photon energy, $\Gamma_1 = \frac{1}{T_1}$ and $\Gamma_2 = \frac{1}{T_2}$ are the diagonal and the off-diagonal relaxation rates, respectively. Here, $T_1 = 1$ ps and $T_2 = 0.2$ ps.

2.2.3. Second Harmonic Generation

Within a three-level system approach, the SHG coefficient is given by [23–25]:

$$\chi_{2\omega}^{(2)} = \frac{e^3 N}{\varepsilon_0} \frac{\mu_{12}\mu_{23}\mu_{31}}{(E_{31} - 2\hbar\omega + i\hbar\Gamma_0)(E_{21} - \hbar\omega + i\hbar\Gamma_0)} \tag{14}$$

where ε_0 is the vacuum permittivity, $N = 10^{22}$ m^{-3} is the volume density of carriers in the system, $\mu_{12}\mu_{23}\mu_{31}$ is the geometric factor in which ($\mu_{ij} = \langle\psi_i|z|\psi_j\rangle, j = 1, 2, 3$) is the off-diagonal matrix element, $E_{31} = E_3 - E_1$ and $E_{21} = E_2 - E_1$ are the transition energies, and $\Gamma_0 = \frac{1}{T_0}$, where $T_0 = 0.14$ ps is the damping term associated with the lifetime of the electron involved in the transitions.

3. Validation

To validate our numerical code based on the FDM, we have considered an ideal cubic QD system of dimensions $L_xL_yL_z = 40$ nm^3 at zero confining potential (Figure 1). Here, the QD is embedded in an infinite barrier potential, defined as $V = 1$ eV. The electron effective mass of carriers is $m^* = 0.029m_0$, as it is of InAs at room temperature $T = 300$ K and zero pressure.

Although this geometry is not practical, it provides a good starting point for any geometrical problem, since the FDM should be useful enough to solve any confinement potential. Also, the Schrödinger equation solution of the considered system, such as the eigenenergy is immediate and is given by:

$$E_{n_x,n_y,n_z} = \frac{\hbar^2\pi^2}{2m^*}\left[\left(\frac{n_x}{L_x}\right)^2 + \left(\frac{n_y}{L_y}\right)^2 + \left(\frac{n_z}{L_z}\right)^2\right]; n_x, n_y, n_z = 1, 2, 3... \tag{15}$$

According to the above analytical expression, the eigenvalues of the first, second, and third-lowest levels are $E_{1\_Analy} = 24.312$ meV, $E_{2\_Analy} = 48.624$ meV, and $E_{3\_Analy} = 72.936$, respectively. The key point of the FDM is the choice of mesh size Δ. Therefore, in Figure 2, we have plotted the energy levels of the two lowest confined states as a function of the grid point number. In the vicinity of $N_z = 64$ nodes, the numerical solutions are $E_{1\_Num} = 24.311$ meV, $E_{2\_Num} = 48.617$ meV and $E_{3\_Num} = 72.923$ meV. Consequently, the numerical results obtained via our code converge to those calculated analytically.

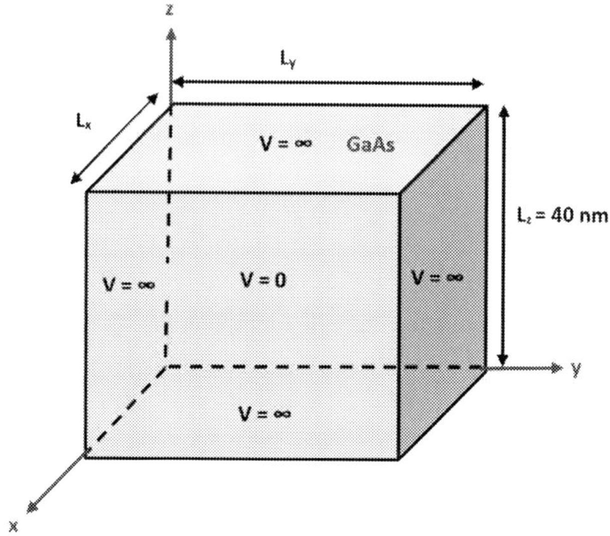

Figure 1. Ideal cubic InAs QD system of dimensions $L_xL_yL_z = 40\ \mathrm{nm}^3$ at zero confining potential, and embedded in an infinite barrier potential.

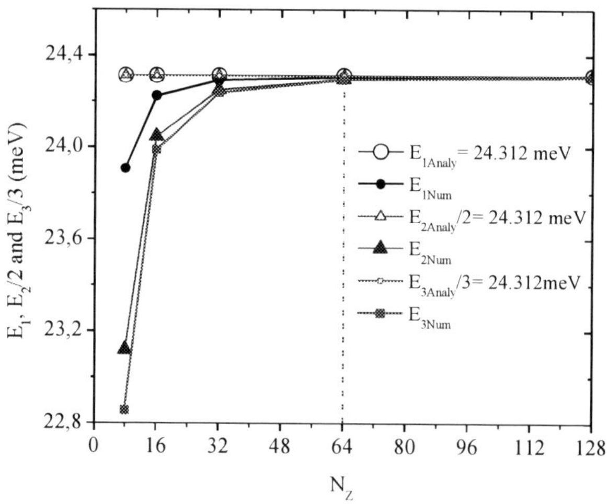

Figure 2. Variation of the first, second, and third-lowest energy levels versus N_z.

To measure the rate of convergence, the absolute error between analytical and numerical solutions versus N_z is illustrated in Figure 3. According to this figure, there is a good agreement between the numerical and analytical results at the starting point of N_z = 64 nodes. To guarantee convergence and reasonable accuracy for all the remaining simulations, we have adopted N_z = 128 nodes. This choice corresponds to a spatial mesh size Δ = 3.15 Å and an error of about 10^{-3} meV.

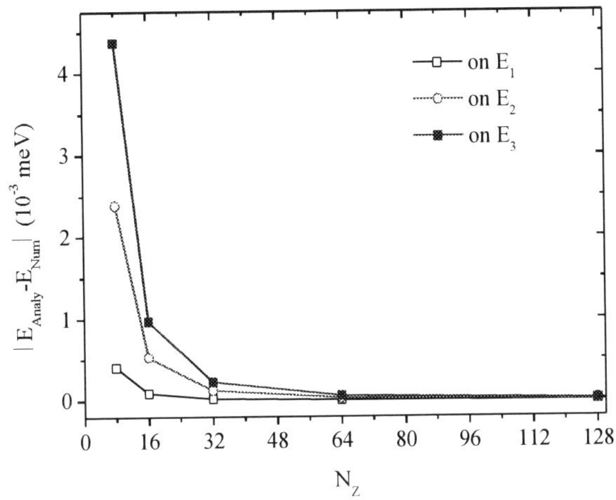

Figure 3. Variation of the absolute error versus N_z.

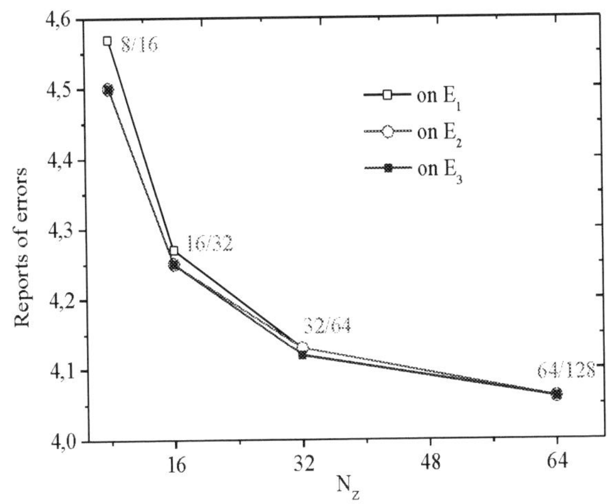

Figure 4. Variation of the reports of errors versus N_z.

Also, the report on errors versus N_z remains in the vicinity of 4, as illustrated in Figure 4. According to Eq. (5), the FDM is of second-order.

Consequently, an implicit FDM of a second order will be used in this work for all the remaining simulations.

4. Single Quantum Dot

According to the literature, QDs have been considered as pyramidal, elliptical and dome shaped. However, based on transmission electron microscopy (TEM) results and atomic force microscopy (AFM) images, the QD shape was considered lens-shaped. Therefore, to generalize our numerical code based on the FDM, we have carried out our calculations for a single lens shaped InAs QD with a radius $R = 7$ nm and a variable height h (Figure 5).

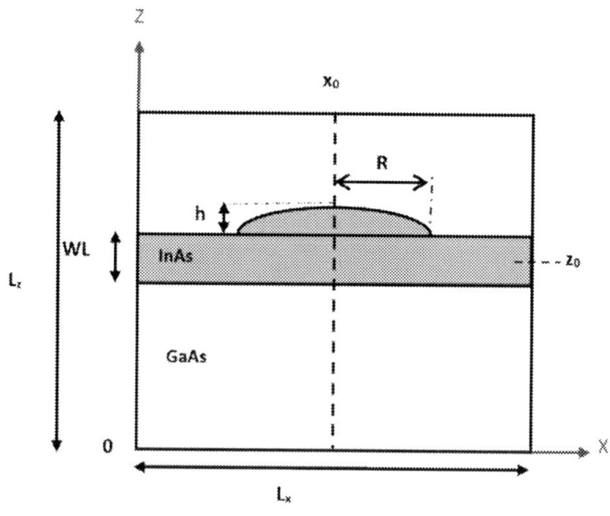

Figure 5. Schematic cross-section at $y_0 = L_y/2$ of a single lens-shaped InAs QD.

The SQD with radius $R = 7$ nm and variable height h is grown on a wetting layer with a thickness of $WL = 0.5$ nm at the centre of a GaAs box. This structure is considered by *M. Shahzadeh et al.* [26] and calculations are carried out exploiting the finite element method (FEM). In these simulations, the electron effective mass was set as $m^* = 0.04m_0$ for InAs and $m^* = 0.067m_0$ for GaAs. The confinement potential was fixed to zero for InAs and to $V = 500$ meV for GaAs. Figure 6 shows a cross-section at y_0 of the confinement potential for the structure under investigation.

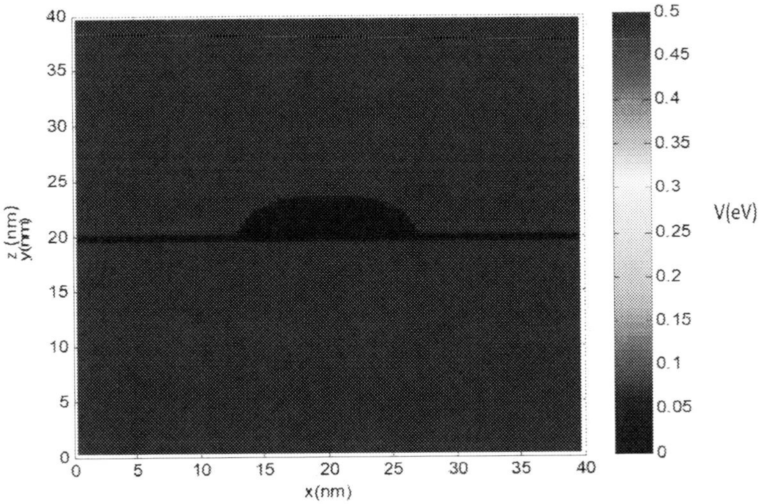

Figure 6. Cross-section at $y_0 = L_y/2$ of the confinement potential for the SQD.

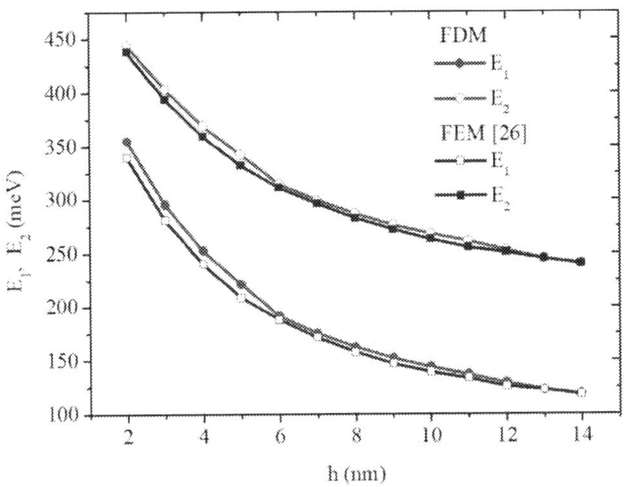

Figure 7. Variation of the first lowest energy levels as a function of the QD height (R = 7 nm) [27].

Figure 7 displays the first two energy levels E_1 and E_2 as a function of QD height. It is seen that the obtained results are in fair agreement. Furthermore, the first lowest energy levels decreased with increasing the QD height. This is due to the confinement reduction of electrons in the QD.

In addition, during these simulations, we have taken the opportunity to show the wetting layer effect on the energy levels. In this stage, we have considered a single lens-shaped QD with height $h = 5$ nm. Using the FDM, the transition energy between the ground and first excited states as a function of radius width and for two WL widths are illustrated in Figure 8.

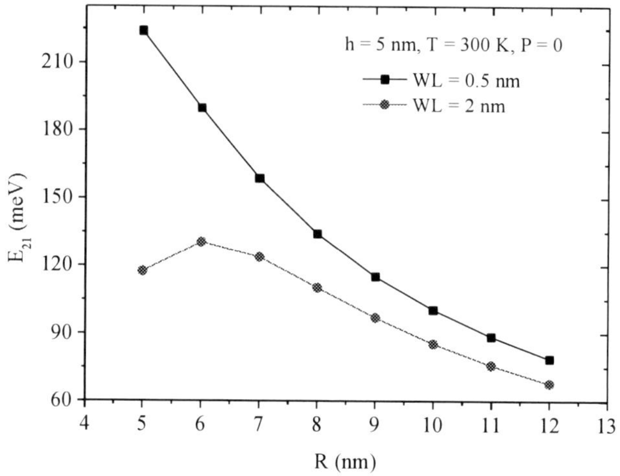

Figure 8. Transition energy versus radius for two wetting layer widths.

The transition energy displayed a red shift (lower energies) with an increase in R. This is attributed to the fact that the potential height changed the quantum confinement effect of the structure under simulation. So, the energy difference decreases when R increases. Also, for a constant height, when the wetting layer width was increased from 0.5 to 2 nm, the transition energy decreased. This decrease is due to the smaller confinement of electrons. Moreover, a noticeable difference between the two cases $WL = 0.5$ nm and $WL = 2$ nm was observed up to approximately $R = 6$ nm before starting exponential variations. This is attributed to the location of the electron envelope function in the wetting layer for small QDs. Thus, the wetting layer imposes its effect only in a small radius. To further explain this effect, we have plotted the probabilities of finding an electron inside the QD region and in the wetting layer at a fixed thickness $WL = 0.5$ nm, according to experimental reports. Hence, the probability of finding an electron inside the QD or in the WL is given by:

$$\frac{\int_{QD/WL} |\psi_i|^2 d^3r}{\int_{system} |\psi_i|^2 d^3r} \times 100 \tag{16}$$

The obtained results are shown in Figure 9. According to this figure, the wetting layer imposes its effect on a smaller QD.

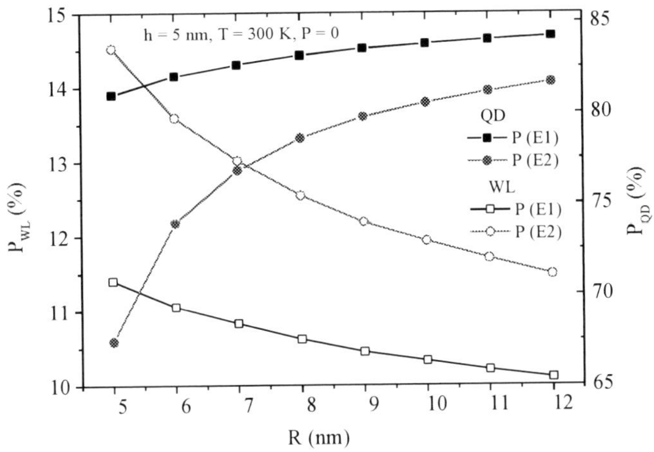

Figure 9. Probabilities of finding an electron in the wetting layer and inside QD versus QD radius for the two first lowest energy levels.

Moreover, when the QD radius was increased, the probability of finding electrons in the wetting layer decreased but increased in the QD region for both energy states. Therefore, one can deduce that the envelope function extended in InAs wetting layer and GaAs matrix moves inside the QD. This effect has been illustrated by a two- dimensional distribution of the ground and first excited states envelope functions for a small semi-spherical QD ($h = R = 5$ nm) as shown in Figure 10. It is seen, for $WL = 0.5$ nm that the ground and first excited states are located inside the QD. However, when the wetting layer was increased, both envelope functions moved towards the wetting layer and GaAs matrix. From these results, we conclude that the NLOP and resonant behavior strongly depend on the QD dimensions and the wetting layer effect. Thus, this energy shift would be manifested in the NOR coefficient already discussed.

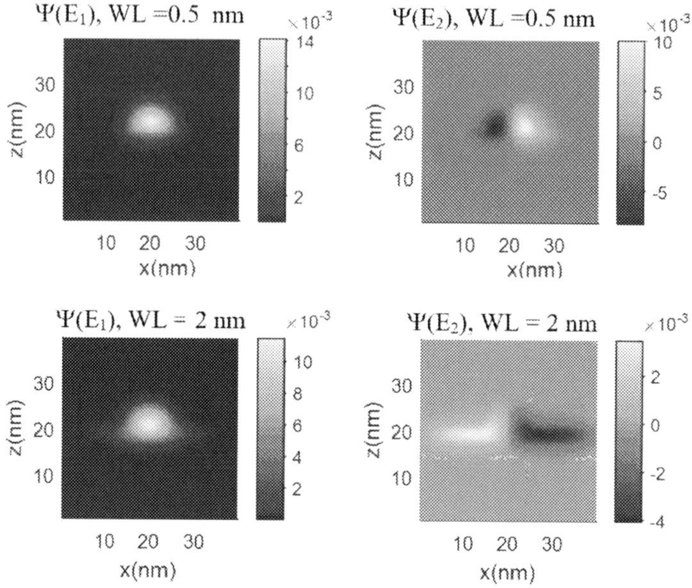

Figure 10. Two-dimensional distribution of the ground and the first excited states envelope wave functions (WL = 0.5 nm and WL = 2 nm in the up and down figures, respectively).

To further understand the behavior of the transition energy and the energy eigenvalues, we look at the advantage of the QD height effect. The obtained results are illustrated in Figure 11.

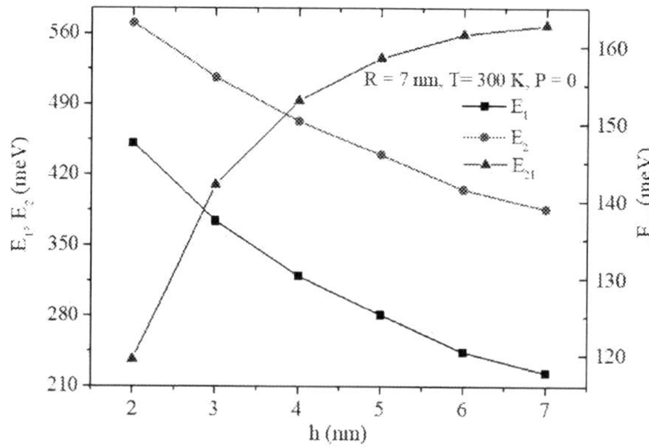

Figure 11. Variation of the two first lowest energy levels and the transition energy versus QD height.

Our simulation was performed for $R = 7$ nm, $T = 300$ K, $WL = 0.5$ nm and at zero pressure. Figure 9 shows that the energy eigenvalues decreased with increasing QD height. This is due to a reduction of the carrier confinement inside the QD region. Hence, a blue-shift (higher energies) is obtained. The results revealed that the transition energy can be red- or blue-shifted energies depending on the structural dimensions and the wetting layer effects. Therefore, these effects can be used as tuning parameters to achieve a large NOR with adjusted resonant energy. Moreover, in addition to the wetting layer and structural dimensions, in this work, pressure and temperature effects are also introduced. As an example, in Figure 12, we plotted the NOR coefficient as a function of the incident photon energy for different pressure and temperature values.

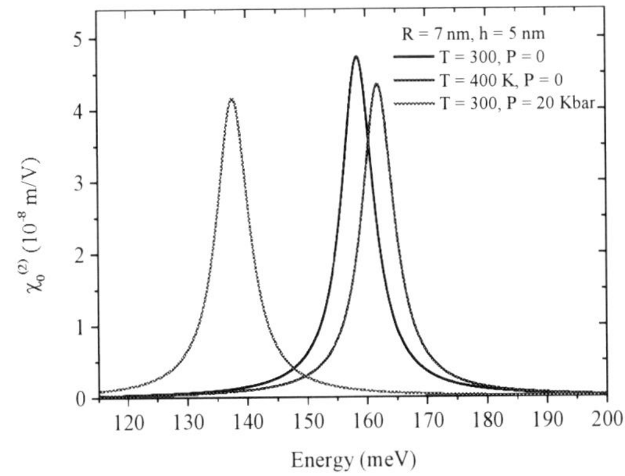

Figure 12. NOR coefficient as a function of the incident photon energy for different pressure and temperature values.

We observe that a red-shift was achieved and the NOR magnitude increased with increasing pressure. This is due to an increase of the confining potential, followed by a decrease in the separation energy between the ground and the first excited energies. Nevertheless, a blue-shift was achieved with increasing temperature. The obtained results allow designing an SQD coupled to its wetting layer and based on the NOR coefficient with adjusted maximum values and resonant energies. To summarize, by using the lens-shaped QD, wetting layer, pressure, and temperature effects, the NOR peak, and the corresponding resonant energy can be adjusted and controlled by increasing pressure or temperature. Hence, a red- or a blue shift can be achieved.

5. Laterally Coupled Quantum Dots

Recently, from the AFM and TEM results, the coupling between semiconductor QDs can be realized either by vertical stacking along the growth axis (z-axis) and by a lateral distribution of the QDs in the same plane perpendicular to this axis. Hence, the QDs are coupled to each other both laterally and vertically. Therefore, the electronic and optical properties are different from those of an SQD. Recently, more attention has been paid to the coupled QDs and different shapes are investigated in literature to adjust and match the experimental results. However, to our knowledge, the lateral coupling effect of two neighboring QDs received less attention. Thus, in the present work, we focus on studying the lateral coupling effect of QDs on the NOR coefficient, the transition energy, and the intra-band transition lifetime of carriers given by [28]

$$\tau = 3c^3 h^4 \varepsilon_0 \Big/ 16\pi^3 n_r E_{21}^3 \left|\mu_{12}\right|^2 \tag{17}$$

Also, the wetting layer pressure and temperature effects were introduced together. Figure 13 displays a schematic cross-section of the structure under investigation.

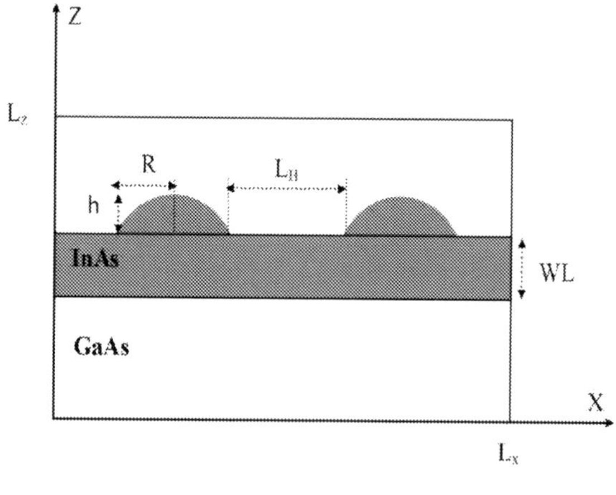

Figure 13. Schematic cross-section at $y_0 = L_y/2$ of two laterally coupled lens-shaped QDs coupled to a wetting layer.

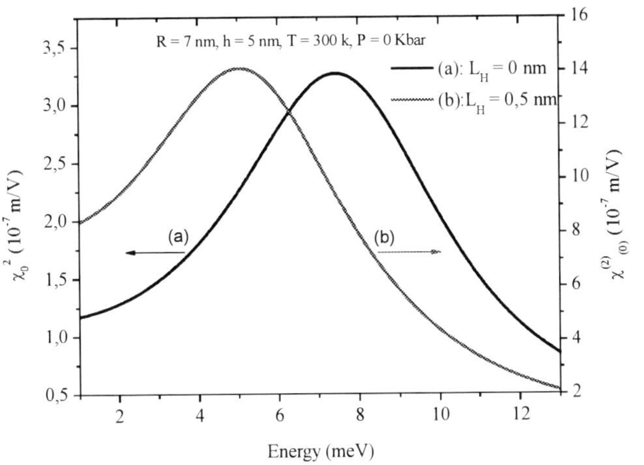

Figure 14. Spacer width effect on the NOR as a function of the incident photon energy for a laterally coupled QDs.

The variation of the NOR of such nanostructure versus the incident photon energy is shown in Figure 14. From this figure, we conclude that the spacer layer is responsible for tunneling the envelope functions of an electron from one QD to its adjacent. So, an increase of the spacer width (distance between neighboring QDs increase to about 6 nm), cut the correlation between QDs and the system behaved as two independent QDs. In this case, the NOR magnitude increased and was followed by a red-shift of the corresponding resonant energy. Moreover, we have found that the transition lifetime of the order of $\tau = 54.4\mu s$ obtained for small spacer increased drastically to $\tau = 6$ s for decoupled QDs. This phenomenon can be explained as follows: increasing the spacer layer from zero to $L_H = 6$ nm, the coupled QDs system approximated to be a single one. So, the ground (E_1) and first excited (E_2) states of the coupled QDs tended to the ground state of an SQD, and the corresponding envelope functions vanished in the wetting layer connecting both the QDs. As a result, the transition energy (E_{21}) and the dipole moment (μ_{12}) tended to zero. Consequently, according to Eq.(13), the product $E_{21}^3|\mu_{12}|^2$tended to zero. Hence, the intra-band transition lifetime increased significantly. Besides, at zero spacer layer, the structure under investigation had a resonant behavior in the THz range (1.2–2.88 THz), as depicted in the same figure. Also, the optimum value of the NOR coefficient $\chi_{0\ Max}^{(2)}$ was achieved at different resonant energies (ω_{res}) in the THz domain, with a red or a blue-shift, upon increasing the applied hydrostatic pressure or temperature, respectively, as showed by Figure 15.

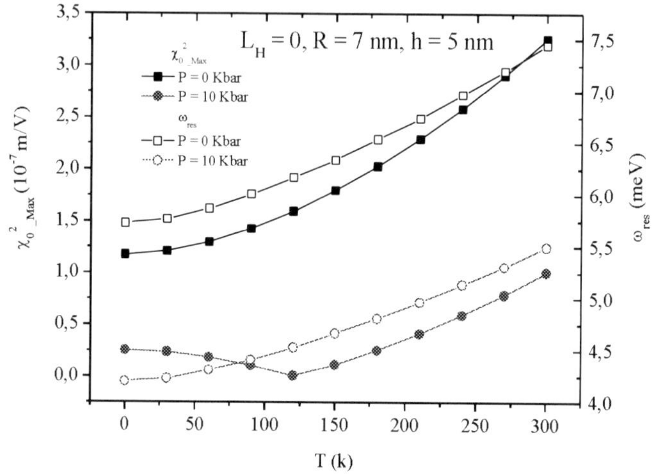

Figure 15. The optimum value of the NOR and the corresponding resonant energy as a function of temperature for different values of pressure for a single QD.

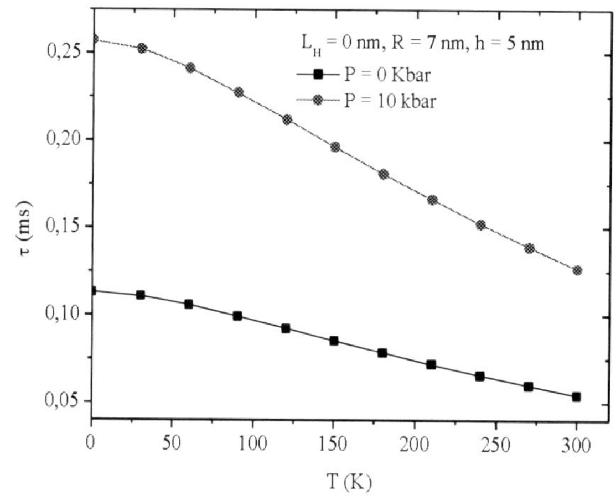

Figure 16. Intra-band transition lifetime as a function of temperature for different values of pressure.

For clarity, Figure 16 presents the intra-band transition lifetime of the structure under investigation as a function of pressure and temperature. At a fixed temperature, it is clear that the transition lifetime increased with the increase in pressure. So, the increase in pressure enlarged the confinement potential and the effective masses of carriers, followed by an increase in the intra-band transition lifetime between the ground and the first excited states.

However, at a fixed pressure, when the temperature was increased, the opposite phenomenon was obtained. So, the intra-band transition lifetime decreased and a blue-shift of the NOR coefficient was achieved, which remained in the THz range. Moreover, at the room temperature and zero pressure, it is seen from Figures 14 and 15, using the laterally coupled effect with zero spacer width reduced the NOR magnitude from $\chi_0^{(2)} = 4.7510^{-8} m/V$ to ~3.25×10^{-7} m/V as compared to an SQD.

6. Vertically Coupled Quantum Dots

In this work, a schematic cross-section of the simulated system is illustrated in Figure 17. According to this figure, the structure under study is made up of two vertically coupled lens-shaped QDs (VCQDs) distributed on a wetting layer. Here, InAs QDs are embedded in a three-dimensional GaAs matrix, where $L_x = L_y = L_z = 40$ nm. Both the stacked lens-shaped InAs QDs are separated by an L_B GaAs buffer. For the VCQDs, the base length was considered to be constant ($2R = 14$ nm) while their heights are $h_1 = 5$ nm and $h_2 = 4$ nm for the lower and upper QDs, respectively.

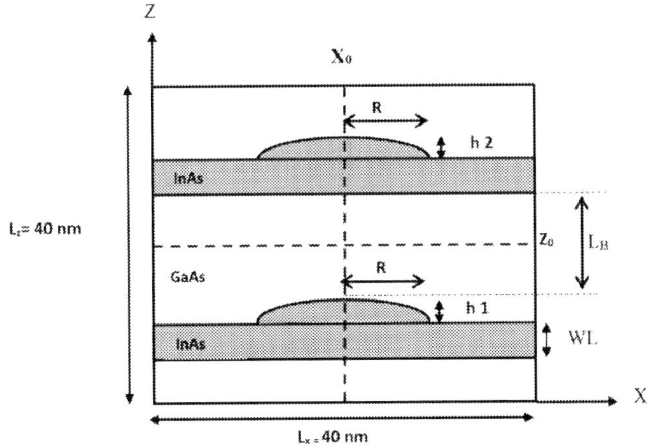

Figure 17. Schematic cross-section at $y_0 = L_y/2$ of two vertically coupled lens-shaped QDs along the growth direction.

In this section, we focused only on the vertical coupling effect between the QDs under hydrostatic pressure, temperature, and the external applied electric field. Also, we search to optimize the SHG by downscaling these external proofs. First, in the calculations of Figures 17–19, the electric field (F) was zero and its effect was introduced later. For an arbitrarily coupled width $L_B = 2.5$ nm, the SHG coefficient versus photon energy is shown in Figure 18.

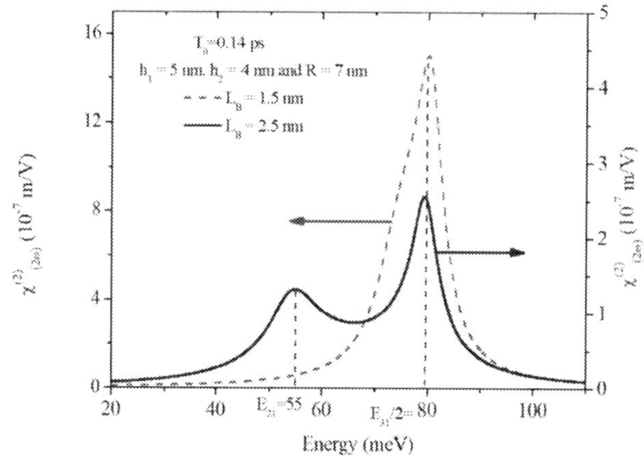

Figure 18. SHG coefficient versus incident photon energy with wetting layer and forLB = 2.5 nm and 1.5 nm (T = 300 K, P = 0 and F = 0) [29].

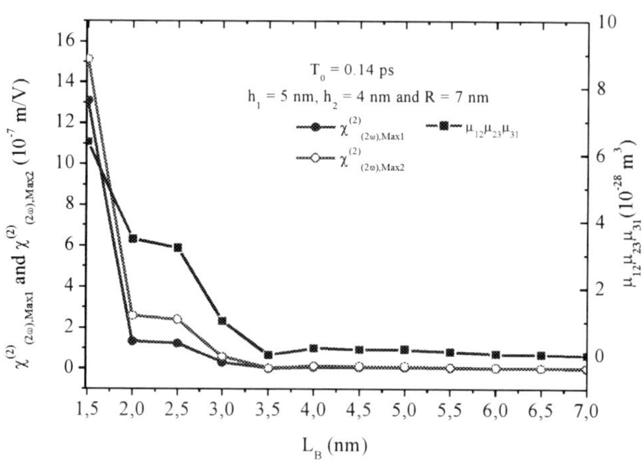

Figure 19. Resonant peak values of the SHG and the geometrical factor versus the coupled width (T = 300 K, P = 0, and F = 0).

According to Eq. (14), the SHG has minor $\chi^{(2)}_{2\omega,Max1} = 1.510^{-7}$ m/V and major $\chi^{(2)}_{2\omega,Max2} = 2.510^{-7}$ m/V peaks located in the vicinity of $\hbar\omega = E_{21} = E_2 - E_1 = 55\,\text{meV}$ and $\hbar\omega = E_{31} = \frac{E_3 - E_1}{2} = 80\,\text{meV}$, respectively.Now, the aim is to investigate the coupled width effect on the resonant peak values and then on the geometrical factor ($\mu_{12}\mu_{23}\mu_{31}$). The obtained results are illustrated in Figure 19. It is seen that the resonant peak values are directly proportional to the geometrical factor, which decreased when increasing L_B. This behavior can be explained as follows: increasing L_B significantly reduces the coupling effect between the vertical QDs. Hence, the overlap between the envelope functions and the dipole matrix element decreases. As a result, the geometrical factor reached its smallest value when increasing the coupled width.

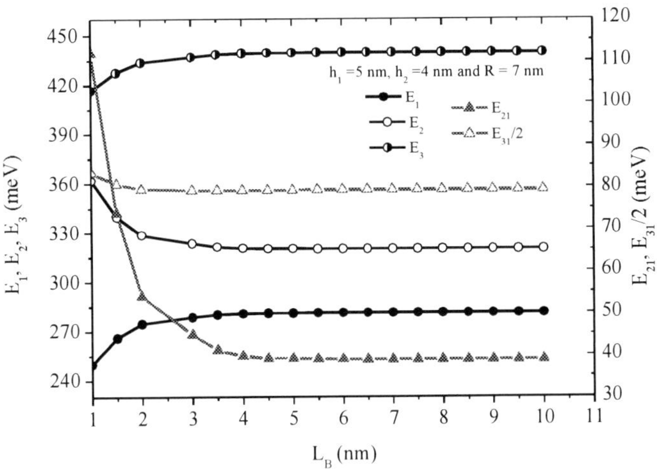

Figure 20. Variation of the first, second, and third-lowest energy levels and the transition energies versus coupled width (T = 300 K, P = 0 and F = 0).

As well, in Figure 20, we investigated the coupled width effect on the first lowest energy levels (E_1, E_2, and E_3) and then on the resonant energies E_{21} and $E_{31}/2$ of the SHG. The resonant peaks underwent a red-shift upon increasing L_B. Thus, decreasing the coupling effect between the QDs led to a reduction of energy confinement. Also, the obtained results showed that the resonant energy E_{21} increased faster than $E_{31}/2$, than they met at the position of length L_B = 1.5 nm with $E_{21} = E_{31}/2$ = 80 meV. That is to say – double- resonance can be realized at this location. According to these results, we can conclude that the largest value of SHG can be obtained by choosing a smaller coupled width between QDs. For all the remaining simulations, we have adopted L_B = 1.5 nm. For this coupled width, the optimized SHG coefficient versus photon energy is shown in the same figure (Figure 18) which shows that an optimum value of about 15 10^{-7} m/V is reached at the resonant energy of about 80 meV.

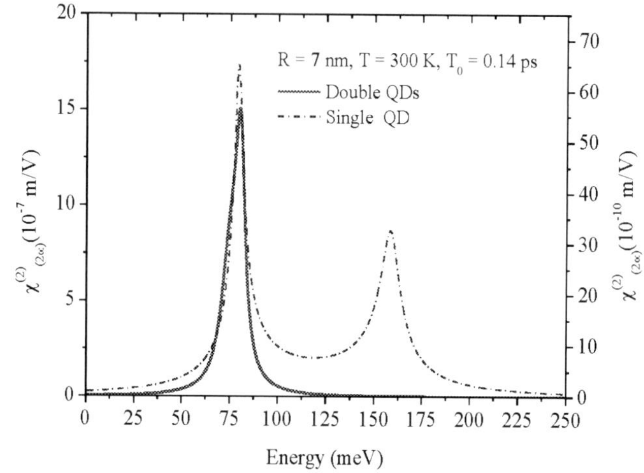

Figure 21. Dependence of the SHG coefficient on the QD number (T = 300 K, P = 0, and *F*= 0).

To better understand the dependence of the SHG coefficient on the number of coupled QDs along the growth direction, we plotted the SHG curves versus the incident photon energy for the coupled QDs, as illustrated in Figure 21. For comparison, the results obtained for the single lower QD are given by the same figure. It is observed, that increasing the QDs number induces an increase in the SHG magnitude followed by a red-shift to lower energies. This behavior of the SHG coefficient could be attributed to the increase of the geometrical factor $\mu_{12}\mu_{23}\mu_{31}$ with increasing the QD number coupled along the growth direction.

For a more accurate analysis, in this work, we have also considered the external electric field effect on the SHG. For this purpose, we have considered an electric field (*F*) applied along the *z*-direction. Figure 22 shows the variation of the minor and major peaks of the SHG coefficient as a function of the value electric field.

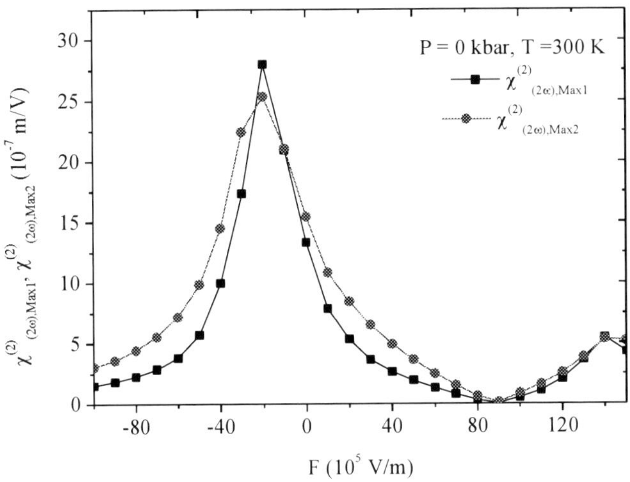

Figure 22. Resonant peak values of the SHG versus electric field (T = 300 K, P = 0 kbar).

We readily see that for F = -20×10^5 V/m, the SHG coefficient reaches an optimum value of about 28×10^{-7} m/V. This negative electric field strengthens the asymmetry of the structure under investigation and, consequently, the degree of overlap between the states falls. Then, at such an electric field, there must be a maximum of the geometrical factor which increases quickly and leads to an increase of the SHG magnitude. Therefore, the electron envelope function can be shifted by applying an electric field along the growth direction. As a result, the electric field induces a deformation of wave functions, and therefore, the corresponding probability densities of the three states involved. Hence, a negative electric field pulls the probability density of the electron upward, while a positive one pushes it downward.

To summarize, the SHG coefficient of the optimized structure (L_B = 1.5 nm, T = 300 K, P = 0 kbar) under different external electric field values are illustrated in Figure 23. From this figure, it can be seen that not only the major resonant peak intensity but also the minor is influenced by the applied electric field. This is due to the dependence of the geometrical factor on the electric field intensity and orientation.

As a summary, the effects of coupled width between QDs, wetting layer, pressure, temperature and the external applied electric field on the SHG coefficient and their corresponding resonant energies were investigated and explored. The obtained results are very important to design VCQDs based on the SHG coefficient with adjusted maximum values and resonant energies. As well, a red- or a blue-shift was achieved, when the external effects were introduced. So, the spacer width (L_B) can be used as a tuning parameter to correlate theoretical and experimental results.

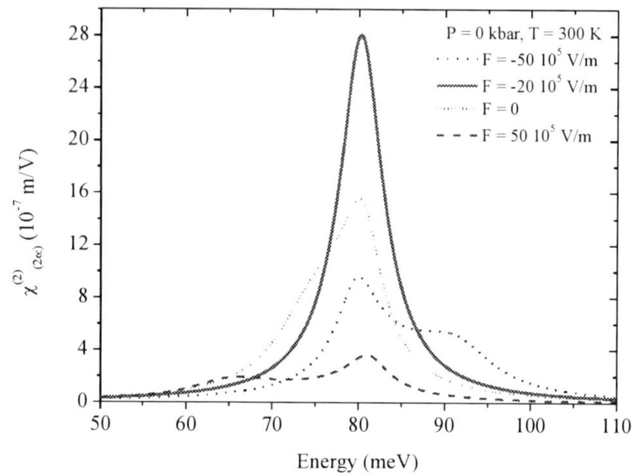

Figure 23. SHG coefficient versus the photon energy for different external electric field values (P = 0 kbar and T= 300 K).

7. Lens-Shaped Core/Shell Quantum Dots

In this section, our rationale is to enhance and adjust the total optical absorption coefficient (α) in InAs/GaAs core/shell QD. According to many theoretical published studies, different shapes of core/shell QD structures have been investigated, such as elliptical and spherical shapes [30, 31]. However, from the TEM results, it is found that the lens-shaped QD is the most accurate model of QDs [32, 33]. Therefore, in this work, we have adopted this actual QD model. A cross-section of the structure under study is illustrated in Figure 24. Hence, after the InAs core formation, the GaAs shell material is deposited and crystallized as the barrier with a great energy gap covering the uncapped core material (InAs). It is noticed, that only the lens-shaped QD is cupped with the shell material. Here, the shell sizes are R_2–R_1 and h_2–h_1 in (the x–o–y) plane and along the (001) direction, respectively. In our investigations, R_2 and R_1 are used as the tuning parameters.

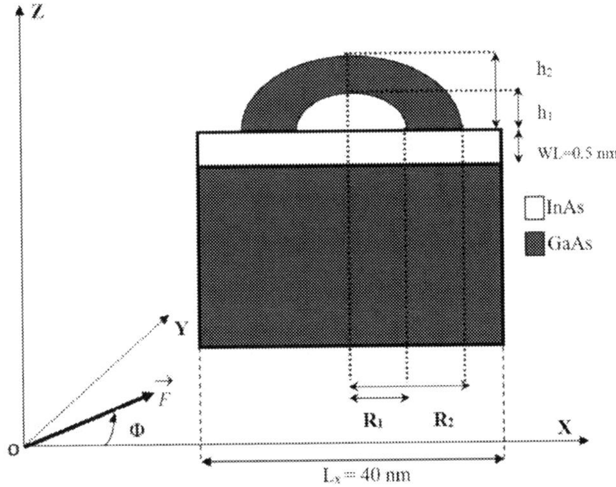

Figure 24. Schematic cross-section at $y_0 = L_y/2$ of the lens-shaped InAs/GaAs core/shell structure.

The essence of this work is to look for a further improvement of the core/shell absorption coefficient under an applied transverse electric field (F). Here, the transverse electric field is oriented along the cross-section (x–o–y) specified by the angle Φ between the positive direction of the x-axis and the electric field direction. The electric field is demonstrated as follows:

$$\vec{F} = F(\cos(\phi)\vec{e_x} + \sin(\phi)\vec{e_y}) \tag{18}$$

Here, the quantity $-eF(x\cos\phi + y\sin\phi)$ is added to the confinement potential, where ($-e$) is the electron charge. This quantity is especially added to the $A(i,j,k)$ coefficients which appear in the finite-difference representation given by Eq. (9).

Figure 25 shows the linear ($\alpha^{(1)}$), nonlinear ($\alpha^{(3)}(\omega, I)$), and the total absorption (α)coefficients as a function of the incident photon energy for three core/shell quantum-dot configurations.

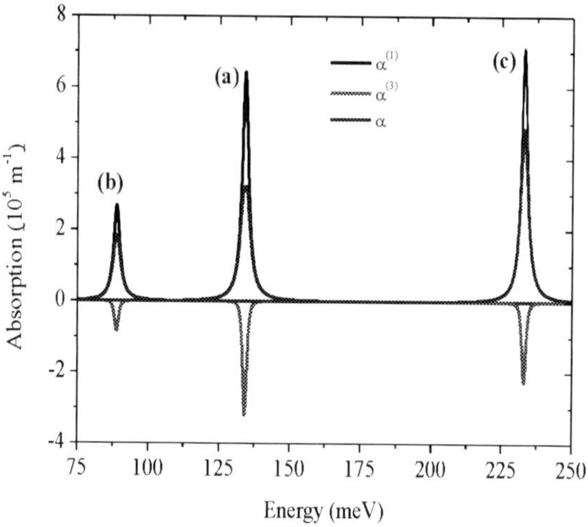

Figure 25. Effects of structural dimensions on the linear, nonlinear and total absorption coefficients versus incident photon energy with h_1 = 5 nm, h_2 = 8 nm, and R_2 = 14 nm. Three structural configurations are: (a) R_1 = 8 nm, (b) R_1 = 11 nm, and (c) $R_1 = h_1$ = 5 nm, $R_2 = h_2$ = 8 nm.

Here, the simulation parameters are set to $P = 0$, $T = 300$ K, $I = 0.05\times10^{10}$ W/m^2 and without an applied external electric field ($F = 0$). As can be seen, for a given structural configuration, $\alpha^{(1)}$ is positive, however, $\alpha^{(3)}$ is negative. As a result, the total absorption coefficient drops. For a matter of comparison, a semi-spherical core/shell QD structure ($h_1 = R_1$ = 5 nm and $h_2 = R_2$ = 8 nm) was also investigated ((c) spectrum). Concerning Figure 3(c), it is noticed that (a) and (b) spectra are red-shifted of about 100 meV and 144 meV, respectively. This is related to the confinement reduction of carriers by increasing the core radius (R_1) or by decreasing the shell width (R_2–R_1). Hence, the control of the core/shell shape is required to adjust the experimentally obtained results. For all the remaining simulations, we have used this structural configuration: h_1 = 5 nm, R_1 = 6 nm, h_2 = 8 nm and R_2 = 14 nm, where the magnitude of the total absorption coefficient reaches $\alpha_{max} = 4.4\times10^5$ m^{-1} value. On the other hand, from Eq. (12), the survey revealed that the third-order non-linear coefficient $|\alpha^{(3)}|$ increased with increasing the incident optical intensity (I); thus, the total absorption coefficient decreased, as illustrated in Figure 26.

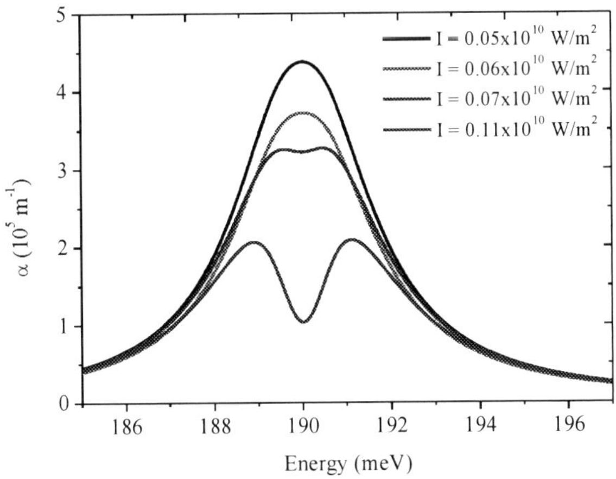

Figure 26. The effect of incident optical intensity on the total absorption coefficient as a function of the incident photon energy (h_1 = 5 nm, R_1 = 6 nm h_2 = 8 nm, and R_2 = 14 nm).

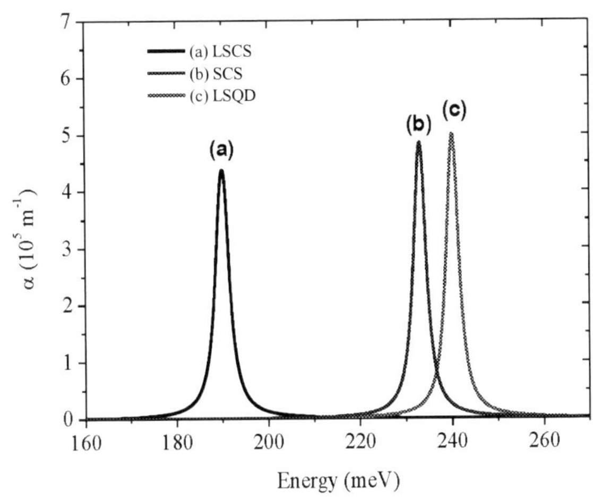

Figure 27. Variation of the total absorption coefficient as a function of incident photon energy: a comparison between: (a) LSCS (h_1 = 5 nm, R_1 = 6 nm, h_2 = 8 nm, R_2 = 14 nm), (b) SSCS(h_1 = R_1 = 5 nm, h_2 = R_2 = 8 nm), and (c) LSQD (h_1 = 5 nm, R_1 = 6 nm) structures.

It is also noticed that the saturation absorption occurs at $I_s = 0.07\times10^{10}$ W/m^2. When the incident optical intensity exceeds this critical value, the non-linear term leads to a collapse at the center and the total absorption coefficient peak splits into two peaks. Consequently, for all the remaining simulations, the incident optical intensity was set to $I = 0.05\times10^{10}$ W/m^2.

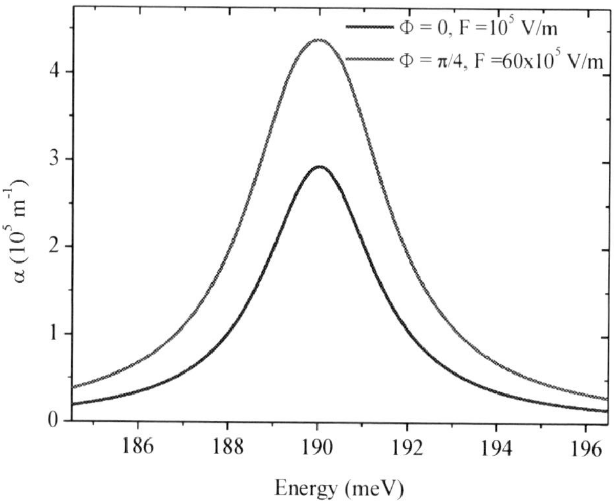

Figure 28. Absorption coefficient as a function of the incident photon energy at low and high electric field intensities [27].

In an attempt to estimate the shell behavior, a comparison between the lens-shaped core/shell quantum dot (LSCS), an entirely capped lens-shaped quantum dot (LSQD), and a semi-spherical core/shell quantum dot (SSCS) systems are displayed in Figure 27. The absorption coefficient of the LSCS displays a 43 meV red-shift owing to the SSCS quantum dot ((a) and (b) spectra). So, the shell material causes a smaller separation between the ground and the first excited states. This can be explained as follows: during the capping process, the redistribution of indium (In) atoms takes place, which reduces the height of QDs and changes their shape and homogeneity. Thus, the capping layer, originally circular, transforms to a lens-shaped QD. For this reason, we have used this realistic model of QDs. Likewise, the shell material causes a 50 meV red-shift owing to the entirely capped SQD ((a) and (c) spectra). Thus, for economic reasons and to lower its cost, one can precisely control the absorption spectrum by downscaling the shell width rather than using QDs entirely covered with GaAs material. However, by the electric field intensity and orientation, one can adjust the total absorption coefficient spectrum in a lens-shaped core/shell quantum dot system. To start with, our simulations are carried out with the pressure, temperature, and the incident optical intensity fixed to $P = 0$, $T = 300$ K and $I = 0.05\times10^{10}$ W/m^2, respectively. In this case, it is found that the heightened absorption coefficient value is achieved when the electric field having low intensity is applied along one side of the lateral cross-section: x- or y-axis ($\Phi = 0$ or $\pi/2$). Furthermore, the heightened absorption coefficient is achieved when the intense electric field is applied along the diagonal direction of the cross-section ($\Phi = \pi/4$). The obtained results are illustrated in Figure 28.

It is seen that the absorption coefficient reaches the optimum value $\alpha_{max} = 4.4\times10^5$ m^{-1} when the electric field is oriented along the diagonal direction of the cross-section (x–o–y) of the structure under investigation. Thus, the magnitude of the absorption coefficient increases by 50%. Accordingly, the designed structure can be used under an intense transverse electric field applied along the diagonal direction of the cross-section of the structure under investigation.

Conclusion

In this chapter, the NLOP in lens-shaped QDs has been studied by using the density matrix approach and the Finite Difference Method in implicit form. Our results show that the magnitude of these properties and the attributed resonant energies are not only affected by the QD height and radius, and the coupled distance between QDs and wetting layer (WL), but also by the external proofs, such as the electric field, pressure, and temperature. With a good choice of these parameters, the resonant energy, and the NLOP magnitude could be adjusted and a red-or/and a blue-shift could be obtained. Moreover, these results may have important consequences for optical studies on semiconductor QDs. Since the size, pressure, and temperature-dependent optical properties of QDs are clearly shown in this work, it suggests many applications, such as adjusting the transition energies of the QD systems coupled to photonic crystals and coupled quantum core/shell. Also, during our simulations, we found that there are more than two confined level states according to the structural dimensions, pressure, and temperature values. This offers an opportunity to study and introduce the effects of these parameters on other nonlinearities, such as the third harmonic generation.

References

[1] Alves, F. D. P., G. Karunasiri, N. Hanson, M. Byloos, H. C. Liu, A. Bezinger, and M. Buchanan, "NIR, MWIR and LWIR quantum well infrared photodetector using interband and intersubband transitions," *Infrared Phys.& Technol.*, vol. 50, no. 2-3, pp. 182–86, 2007.

[2] Darabi, E., V. Ahmadi, and K. Mirabbaszadeh, "Numerical analysis of an optoelectronic integrated device composed of coupled periodic MQW phototransistor and strained-QW laser diode," *Solid-State Electron.*, vol. 50, no. 3, pp. 473–79, 2006.

[3] Chena, B., K.-X. Guob, R.-Z. Wang, Y.-B. Zheng, and B. Li, "Nonlinear Optical Rectification in Asymmetric Double Triangular Quantum Wells," *Eur. Phys. J. B*, vol. 66, no. 2, pp. 227–33, 2008.

[4] Liang, B., P. Wong, N. Pavarelli, J. Tatebayashi, T. Ochalski, G. Huyet, and D. Huffaker, "Lateral interdot carrier transfer in an InAs quantum dot cluster grown on a pyramidal GaAs surface," *Nanotechnol.*, vol. 22, no. 5, p. 055706, 2011.

[5] Sabaeian, M. and A. Khaledi-Nasab, "Size-Dependent Intersubband Optical Properties of Dome-Shaped InAs/GaAs Quantum Dots with Wetting Layer,"*Appl. Opt.*, vol. 51, no. 18, pp. 4176-4185, 2012.

[6] Rezaei, G., M. Vahdani, and B. Vaseghi, "Nonlinear Optical Properties of a Hydrogenic Impurity in an Ellipsoidal Finite Potential Quantum Dot," *Curr. App. Phys.*, vol. 11, no 2, pp. 176–81, 2011.

[7] Vahdani M. and G. Rezaei, "Linear and Nonlinear Optical Properties of a Hydrogenic Donor in Lens-Shaped Quantum Dots," *Phys. Lett. A*, vol. 373, no 34, pp. 3079–3084, 2009.

[8] Cusack, M. A., P. R. Briddon, and M. Jaros, "Electronic Structure of InAs/GaAs Self-Assembled Quantum Dots," *Phys. Rev. B*, vol. 54, no 4, p. R2300, 1996.

[9] Karimi, M. J., A. Keshavarz, G. Rezaei, "Optical Rectification and Second Harmonic Generation of Finite and Infinite Semi-Parabolic Quantum Wells," *J. Comput. Theor. Nan.*, vol. 8, no 7, pp. 1340–45, 2011.

[10] Mohammadi, S. A., R. Khordad, G. Rezaei, "Optical Properties of a Semispherical Quantum Dot Placed at the Center of a Cubic Quantum Box: Optical Rectification, Second and Third-Harmonic Generations," *Phys. E*: Low-Dimens. Syst. Nanostructures, vol. 76, pp. 203–208, 2016.

[11] Karabulut, I., H. Safak, M. Tomak, "Nonlinear Optical Rectification in Asymmetrical Semiparabolic Quantum Wells," *Solid State Commun.*, vol. 135, no. 11-12, pp. 735–38, 2005.

[12] Shahzadeh, M., M. Sabaeian, "The Effects of Wetting Layer on Electronic and Optical Properties of Intersubband P-to-S Transitions in Strained Dome-Shaped InAs/GaAs Quantum Dots," *AIP Adv.*, vol. 4, no 6, p. 067113, 2014.

[13] Ameen T. A. and Y. M. El-Batawy, "Polarization Dependence of Absorption by Bound Electrons in Self-Assembled Quantum Dots," *J. App. Phys.*, vol. 113, no. 19, p. 193102, 2013.

[14] Mora-Ramos, M. E., C. A. Duque, E. Kasapoglu, H. Safri and I. Sökmen, "Linear and Nonlinear Optical Properties in a Semiconductor Quantum Well under Intense

Laser Radiation: Effects of Applied Electromagnetic Fields," *J. Lum.*, vol. 132, no 4, pp. 901–913, 2012.

[15] Vurgaftman, I., J. R. Meyer and, L. R. Ram-Mohan, "Band Parameters for III–V Compound Semiconductors and Their Alloys," *J. Appl. Phys.*, vol. 89, no 11, pp. 5815–5875, 2001.

[16] Safarpour, G., M. Moradi, M. Barati, "Hydrostatic pressure and temperature effects on the electronic energy levels of a spherical quantum dot placed at the center of a nano-wire," Superlattices Microstruct., vol. 52, no. 4, pp. 687–696, 2012.

[17] El Haouari, M., A. Talbi et al., "Linear and Nonlinear Optical Properties of a Single Dopant in Strained AlAs/GaAs Spherical Core/Shell Quantum Dots," *Opt. Commun*, vol. 383, pp. 231–237, 2017.

[18] Kouhi, M. "Study of Nonlinear Optical Absorption in GaN/AlGaN Core–Shell Nanowire," *J. Nanophotonics*, vol. 11, no 1, p. 016012, 2017.

[19] Karimi, M. J., H. Vafaei, "Intense Laser Field Effects on the Linear and Nonlinear Intersubband Optical Properties in a Strained InGaN/GaN Quantum Well," *Phys. B: Condense. Matter*, vol. 452, pp. 131–135, 2014.

[20] Bejan, D., C. Stan, E. C. Niculescu, "Optical Properties of an Elliptic Quantum Ring: Eccentricity and Electric Field Effects," *Opt. Mater.*, vol. 78, pp. 207–19, 2018.

[21] Shen, Y. R. "*The Principles of Nonlinear Optics*," Wiley-Interscience, Hoboken (NJ), (2003).

[22] Hassanabadi, H., G. Liu, and L. Liangliang, "Nonlinear optical rectification and the second harmonic generation in semi-parabolic and semi-inverse squared quantum wells," *Solid State Commun.*, vol. 152, no. 18, pp. 1761–1766, 2012.

[23] Ungan, F., J. C. Martínez-Orozco, R. L. Restrepo, M. E. Mora-Ramos, E. Kasapoglu and C. A. Duque, "Nonlinear Optical Rectification and Second-Harmonic Generation in a Semi-Parabolic Quantum Well under Intense Laser Field: Effects of Electric and Magnetic Fields," *Superlattices Microstruct.*, vol. 81, pp. 26–33, 2015.

[24] Shao, S., K.-X. Guo, Z.-H. Zhang, N. Li, and C. Peng, "Studies on the Second-Harmonic Generations in Cubical Quantum Dots with Applied Electric Field," *Phys. B: Condens. Matter*, vol. 406, no 3, pp. 393–96, 2011.

[25] Hassan, H., G. Liu, and L. Liangliang, "Nonlinear Optical Rectification and the Second-Harmonic Generation in Semi-Parabolic and Semi-Inverse Squared Quantum Wells", *Solid State Commun.*, vol. 152, no 18, pp. 1761–66, 2012.

[26] Shahzadeh, M., M. Sabaeian, "A Comparison between Semi-Spheroid- and Dome-Shaped Quantum Dots Coupled to Wetting Layer," *AIP Adv.*, vol. 4, no 6, p. 067134, 2014.

[27] Choubani, M., H. Maaref, F. Saidi, "Nonlinear Optical Properties of Lens-Shaped Core/Shell Quantum Dots Coupled with a Wetting Layer: Effects of Transverse Electric Field, Pressure, and Temperature," *J. Phys. Chem. Solids.*, vol. 138, p. 109226, 2020.

[28] Yariv, A. "*Quantum Electronics*", Wiley, New York, NY, 1988.

[29] Choubani, M., D. Makhlouf, F. Saidi, and H. Maaref, "Enhancement of the Second Harmonic Generation in a Coupled Lens-Shaped Quantum Dots under Wetting

Layer, Temperature, Pressure, and Electric Field Effects," *Opt. Quantum. Electron*, vol. 52, no 2, pp. 52-66, pp.1-15, 2020.

[30] Taş, H., M. Şahin, "The Electronic Properties of a Core/Shell/Well/Shell Spherical Quantum Dot with and without a Hydrogenic Impurity," *J. Appl. Phys*., vol. 111, no. 8, p. 083702.

[31] Ismail, A. A., A. V. Gholap, Y. A. Abbo "Enhancement of local electric field in core-shell orientation of ellipsoidal metal/dielectric nanoparticles," *Condens. Matter. Phys*, vol. 20, no. 2, p. 23401, 2017.

[32] Chen, J. X., A. Markus, A. Fiore, U. Oesterle, R. P. Stanley et al., "Tuning InAs/GaAs Quantum Dot Properties under Stranski-Krastanov Growth Mode for 1.3 Mm Applications," *J. Appl. Phys*., vol. 91, no 10, p. 6710, 2002.

[33] Bouzaïene, L., L. Sfaxi, M. Baira, H. Maaref, C. Bru-Chevallier, "Power Density and Temperature Dependent Multi-Excited States in InAs/GaAs Quantum Dots," J. *Nanoparticle Res*, vol. 13, no 1, pp. 257–62, 2011.

Chapter 6

Quantum Properties of Light in Codirectional Coupler with Second Harmonic Generation

Mohd Syafiq M. Hanapi[1],
Abdel-Baset M. A. Ibrahim[1,*], **Rafael Julius**[2],
Ahmad M. Alrub[3] and **Pankaj K. Choudhury**[4]
1Faculty of Applied Sciences, Universiti Teknologi MARA (UiTM), Selangor, Malaysia
[2]Faculty of Applied Sciences, Universiti Teknologi MARA (UiTM) Perak, Tapah Campus, Perak, Malaysia
[3]Department of Physics, Al-Hussein Bin Talal University, Ma'an, Jordan
[4]Institute of Microengineering and Nanoelectronics, Universiti Kebangsaan Malaysia, Selangor, Malaysia

Abstract

Second-order nonlinearity is the beneficial source of second harmonic generation (SHG). In this chapter, the generation of squeezed states of light in two-channel coupler with second-order nonlinearity is presented. The investigation is performed theoretically using two mathematical methods, namely the *phase space method* and the *analytical method*. The results reveal that the analytical method outperformed the phase space method in the study of this system. The phase space method required a time-consuming numerical simulation, while the analytical method did not. The detailed mathematical derivation and the graphical result of squeezing can be found throughout this chapter.

Keywords: quantum optics, second harmonic generation, optical coupler, squeezed states

* Corresponding Author's Email: abdelbaset@uitm.edu.my.

In: Second Harmonic Generation
Editors: Abdel-Baset M. A. Ibrahim and Pankaj Kumar Choudhury
ISBN: 978-1-68507-888-1

1. Introduction

To ensure the advancement in future technology, relying on the classical states of light alone is not sufficient, since the classical states have a natural limitation, that is called "standard quantum limit." Due to this limitation, the classical states cannot be applied in various quantum optical applications that require the operation beyond the quantum limit. Therefore, the quantum states, such as squeezing, antibunching, and entanglement, are crucial for future generations, and therefore, intensive study regarding this topic is necessary. For instance, squeezing can be applied in various applications, such as in quantum computing [1], quantum teleportation [2, 3], quantum precision measurement [4, 5], transmission of classical information [6], and many others. All these applications motivate researchers in this field to study the quantum properties of light either theoretically or experimentally.

Nonlinear phenomena are the main criteria in generating the quantum mechanical states. When light interacts with a nonlinear system/device, its polarization (P) is given by

$$P(t)=\varepsilon_0\chi^{(1)}E(t)+\varepsilon_0\chi^{(2)}E^2(t)+\varepsilon_0\chi^{(3)}E^3(t)+\dots \tag{1}$$

where ε_0 is the permittivity of empty space, $\chi^{(n)}$ is the n-th order dielectric susceptibility and $E(t)$ is the electric field intensity. The first term in Eq. (1) is dedicated to the linear interaction, and the rest are the nonlinear terms. Therefore, if light interacts with the linear optical system, the polarization of the light will be dominated by the first term only. As such,

$$P(t)\approx\varepsilon_0\chi^{(1)}E(t) \tag{2}$$

The second and the third terms in Eq. (1) represent the second- and the third-order nonlinear phenomena, respectively. Since the second-order nonlinear phenomena can be the beneficial source for second harmonic generation (SHG), the discussion of this chapter is focused on the generation of quantum states in the optical coupler system made of the second-order nonlinearity material.

Optical coupler is the chosen topic in this chapter, since it has many benefits among the other generator of quantum states. Here, a two-channel symmetric coupler with the second-order nonlinearity, as shown in Figure 1, is considered. As can be seen in this figure, the examined system has a simple structure, where it simply contains two-channel waveguides that are made of the material that shows the second-order nonlinearity. Due to its simple structure, it can be easily controlled by changing the separation distance between the waveguides, and introducing the contra-directional propagation, in which one mode is projected in an opposite direction. Moreover, it can be easily utilized in various applications, thereby making it experimentally realizable. As mentioned in the literature, optical coupler can be applied to build quantum circuits [7], photonic chips [8], and optical OR and NAND gates [9], where all of these are the essential component in a quantum computer. Therefore, if quantum states can be produced in optical coupler, the realization of a quantum computer will no longer be a big issue.

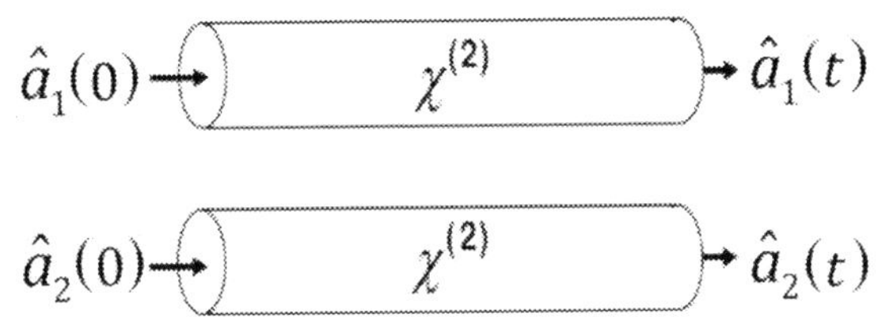

Figure 1. Schematic diagram of the codirectional two-channel symmetric coupler with second-order nonlinearity.

In this chapter, the system will be investigated theoretically using two different mathematical methods – the *phase space method* and the *analytical method*. The phase space method is based on the Schrödinger picture where the quantum operator is assumed to be constant, while the state vector varies with time. As will be seen later, the state vector will be represented by the density matrix $\hat{\rho}$. As such, the system dynamics can be identified by solving the quantum master equation of $\hat{\rho}$. This attempt will require the Stochastic process that involves a lot of computational processes, hence consuming a lot of time. Despite having a time-consuming issue, the phase space method is capable to produce a beneficial result, as it is successfully applied in various research work. Oppositely, the analytical method employs the Heisenberg picture of quantum mechanics, in which the kinetics of the system is represented by the time-dependent annihilation-mode-operator, through the Heisenberg equation of motion. Meanwhile, the state vector is a time-dependent quantity. Using this method, the system's solutions are derived without implementing the Stochastic process, thereby requiring least computational work. For the past several years, the analytical method became well-known to study the quantum states in various systems including the optomechanical-like systems [10, 11], Raman process [12–15], Bose-Einstein condensates [16] and light-semiconductor interaction [17]. For the coupler system, the analytical method has been applied to study the two-channel Kerr coupler [18, 19], two-channel asymmetric coupler with the second-order nonlinearity [20–23]. In these, one waveguide is made of nonlinear material, while the other one is linear, and the two-channel symmetric coupler with the second-order nonlinearity. However, for the symmetrical two-channel coupler with the second-order nonlinearity, only the investigation of quantum Zeno and anti-Zeno effects has been done previously [21]. As such, in this chapter, the focus will be on another important quantum state which is squeezing. A detailed explanation of the mathematical derivation will be discussed in the following section.

2. Mathematical Formulation

The considered system can mainly be described using the following Hamiltonian operator:

$$\hat{H} = \hbar\Big[\omega_1 \hat{a}_1^{\dagger}\hat{a}_1 + \omega_2 \hat{a}_2^{\dagger}\hat{a}_2 + (2\omega_1)\hat{b}_1^{\dagger}\hat{b}_1 + (2\omega_2)\hat{b}_2^{\dagger}\hat{b}_2 + \frac{ig}{2}\left(\hat{a}_1^{\dagger 2}\hat{b}_1 + \hat{a}_2^{\dagger 2}\hat{b}_2 - \text{h.c.}\right) + \kappa\left(\hat{a}_1^{\dagger}\hat{a}_2 + \text{h.c.}\right)\Big] \tag{3}$$

In the above equation, $\hbar$ is the reduced Planck constant, $\hat{a}_j^{\dagger}(\hat{a}_j)$ and $\hat{b}_j^{\dagger}(\hat{b}_j)$ are the creation and annihilation operators of the pump mode and the second harmonic mode, respectively, and h.c. is the abbreviation for 'Hermitian conjugate'. The two terms at the beginning represent the free-propagation of light-mode in the system, where ω_j

third and the fourth terms describe the propagation of the second harmonic mode. The fifth term signifies the nonlinear coupling between the pump mode and the second harmonic mode, where the controlling parameter g, is called the 'nonlinear coupling constant'. The last term modelled the linear mode coupling between the pump modes, in which this interaction arises when the energy is exchanged through the transference of evanescent waves. The controlling parameter κ is called the 'linear coupling constant'.

2.1. Phase Space Method for Codirectional System

The phase space formulation is well-known in quantum optics literature because it provides logical links between quantum mechanics and classical statistical mechanics. This approach was widely utilized to explore numerous aspects of nonclassical light created in several kinds of quantum couplers, as we should point out to the reader [24-33]. For the two-channel system under consideration, the Hamiltonian Eq. (3) is substituted into the following Liouville-von Neumann equation:

$$\frac{\partial\hat{\rho}}{\partial t}=\frac{i}{\hbar}\left[\hat{\rho},\hat{H}\right] \tag{4}$$

to obtain the following quantum master equation for the entire density matrix:

$$\begin{aligned}\frac{\partial\hat{\rho}}{\partial t}&=i(2\omega_1)\left(\hat{\rho}\hat{b}_1^{\dagger}\hat{b}_1-\hat{b}_1^{\dagger}\hat{b}_1\hat{\rho}\right)+i(2\omega_2)\left(\hat{\rho}\hat{b}_2^{\dagger}\hat{b}_2-\hat{b}_2^{\dagger}\hat{b}_2\hat{\rho}\right)\\&+i\omega_1\left(\hat{\rho}\hat{a}_1^{\dagger}\hat{a}_1-\hat{a}_1^{\dagger}\hat{a}_1\hat{\rho}\right)+i\omega_2\left(\hat{\rho}\hat{a}_2^{\dagger}\hat{a}_2-\hat{a}_2^{\dagger}\hat{a}_2\hat{\rho}\right)\\&+\frac{g}{2}\left(\hat{a}_1^{\dagger 2}\hat{b}_1\hat{\rho}-\hat{\rho}\hat{a}_1^{\dagger 2}\hat{b}_1+\hat{a}_2^{\dagger 2}\hat{b}_2\hat{\rho}-\hat{\rho}\hat{a}_2^{\dagger 2}\hat{b}_2\right.\\&\left.+\hat{\rho}\hat{a}_1^{2}\hat{b}_1^{\dagger}-\hat{a}_1^{2}\hat{b}_1^{\dagger}\hat{\rho}+\hat{\rho}\hat{a}_2^{2}\hat{b}_2^{\dagger}-\hat{a}_2^{2}\hat{b}_2^{\dagger}\hat{\rho}\right)\\&+i\kappa\left(\hat{\rho}\hat{a}_1^{\dagger}\hat{a}_2-\hat{a}_1^{\dagger}\hat{a}_2\hat{\rho}+\hat{\rho}\hat{a}_1\hat{a}_2^{\dagger}-\hat{a}_1\hat{a}_2^{\dagger}\hat{\rho}\right)\end{aligned} \tag{5}$$

In the former equation, $\hat{\rho}$ is the density matrix/operator and i is the imaginary number. This equation is not easy to solve. Therefore, it is useful to convert it into the classical form. This conversion can be achieved using the quantum-classical correspondence in the positive-P representation. This correspondence is given by

$$\hat{a}^{\dagger}\hat{\rho}\leftrightarrow\left(\beta-\frac{\partial}{\partial\alpha}\right)P \tag{6}$$

$$\hat{a}\hat{\rho}\leftrightarrow\alpha P \tag{7}$$

$$\hat{\rho}\hat{a} \leftrightarrow \left(\alpha - \frac{\partial}{\partial \beta} \right) P \tag{8}$$

$$\hat{\rho}\hat{a}^{\dagger} \leftrightarrow \beta P \tag{9}$$

where α is the classical equivalent of $\hat{a}$, and β is the classical equivalent of $\hat{a}^{\dagger}$, but β is not the complex conjugate of α. Applying this correspondence in Eq. (5), the following Fokker Planck equation can be derived

$$\begin{aligned}\frac{\partial P}{\partial t} = \Bigg[& \left(i\omega_1\alpha_1 - g\bar{\alpha}_1\beta_1 + ik\alpha_2\right)\frac{\partial}{\partial \alpha_1} + \left(\frac{g}{2}\alpha_1^2 + i(2\omega_1)\bar{\alpha}_1\right)\frac{\partial}{\partial \bar{\alpha}_1} \\ & -\left(i\omega_1\beta_1 + g\alpha_1\bar{\beta}_1 + ik\beta_2\right)\frac{\partial}{\partial \beta_1} + \left(\frac{g}{2}\beta_1^2 - i(2\omega_1)\bar{\beta}_1\right)\frac{\partial}{\partial \bar{\beta}_1} \\ & +\left(i\omega_2\alpha_2 - g\bar{\alpha}_2\beta_2 + ik\alpha_1\right)\frac{\partial}{\partial \alpha_2} + \left(\frac{g}{2}\alpha_2^2 + i(2\omega_2)\bar{\alpha}_2\right)\frac{\partial}{\partial \bar{\alpha}_2} \\ & -\left(i\omega_2\beta_2 + g\alpha_2\bar{\beta}_2 + ik\beta_1\right)\frac{\partial}{\partial \beta_2} + \left(\frac{g}{2}\beta_2^2 - i(2\omega_2)\bar{\beta}_2\right)\frac{\partial}{\partial \bar{\beta}_2} \\ & +\frac{1}{2}\left(g\bar{\alpha}_1\frac{\partial^2}{\partial {\alpha_1}^2} + g\bar{\alpha}_2\frac{\partial^2}{\partial {\alpha_2}^2} + g\bar{\beta}_1\frac{\partial^2}{\partial {\beta_1}^2} + g\bar{\beta}_2\frac{\partial^2}{\partial {\beta_2}^2} \right)\Bigg] P \end{aligned} \tag{10}$$

In the above equation, $\bar{\alpha}$ and $\bar{\beta}$ are the classical equivalent of the second harmonic operator $\hat{b}$ and $\hat{b}^{\dagger}$, respectively. This equation describes the temporal variation of the quasiprobability function (P). Using this function, one can determine the state of a quantum system. For example, the quantum states that possess a classical analogue is represented by the positive value of P. Meanwhile, for the states that have no classical analogue, P will violate the first probability axiom by taking a negative value; such states are classified as a nonclassical state. Equation (10) can be solved using the Stochastic method, where the equivalent set of Stochastic differential equations need to be mapped using the Ito calculus rule. As a result, the following stochastic systems are obtained:

$$\frac{d\alpha_1}{dt} = -i\omega_1\alpha_1 + g\bar{\alpha}_1\beta_1 - i\kappa\alpha_2 + \sqrt{g\bar{\alpha}_1}\eta_1(t) \tag{11}$$

$$\frac{d\beta_1}{dt} = i\omega_1\beta_1 + g\alpha_1\bar{\beta}_1 + i\kappa\beta_2 + \sqrt{g\bar{\beta}_1}\eta_2(t) \tag{12}$$

$$\frac{d\alpha_2}{dt}=-i\omega_2\alpha_2+g\bar{\alpha}_2\beta_2-i\kappa\alpha_1+\sqrt{g\bar{\alpha}_2}\eta_3(t) \tag{13}$$

$$\frac{d\beta_2}{dt}=i\omega_2\beta_2+g\alpha_2\bar{\beta}_2+i\kappa\beta_1+\sqrt{g\bar{\beta}_2}\eta_4(t) \tag{14}$$

$$\frac{d\alpha_3}{dt}=-i(2\omega_1)\bar{\alpha}_1-\frac{g}{2}\alpha_1^2 \tag{15}$$

$$\frac{d\bar{\beta}_1}{dt}=i(2\omega_1)\bar{\beta}_1-\frac{g}{2}\beta_1^2 \tag{16}$$

$$\frac{d\bar{\alpha}_2}{dt}=-i(2\omega_2)\bar{\alpha}_2-\frac{g}{2}\alpha_2^2 \tag{17}$$

$$\frac{d\bar{\beta}_2}{dt}=i(2\omega_2)\bar{\beta}_2-\frac{g}{2}\beta_2^2 \tag{18}$$

The parameters η_i, $i \in 1,2,3,4$ represent the white Gaussian noise with zero expectation value $\langle\eta_i(t)\rangle = 1$. These equations can be solved numerically using the fourth-order Runge Kutta method. However, the solutions need to be averaged over the thousands of trajectories to obtain the stable result.

2.2. Analytical Method for Codirectional System

In the analytical method, the Hamiltonian operator in Eq. (1) is further described using the Heisenberg equation of motion which, in general, is expressed as follows:

$$\frac{d\hat{a}}{dt}=\frac{i}{\hbar}\left[\hat{H},\hat{a}\right] \tag{19}$$

By substituting Eq. (3) into Eq. (19), the following equations are derived:

$$\frac{d\hat{a}_1}{dt}=-i\omega_1\hat{a}_1+g\hat{a}_1^{+}\hat{b}_1-i\kappa\hat{a}_2 \tag{20}$$

$$\frac{d\hat{a}_2}{dt} = -i\omega_2\hat{a}_2 + g\hat{a}_2^{\dagger}\hat{b}_2 - i\kappa\hat{a}_1 \tag{21}$$

$$\frac{d\hat{b}_1}{dt} = -i(2\omega_1)\hat{b}_1 - \frac{g}{2}\hat{a}_1^{\,2} \tag{22}$$

$$\frac{d\hat{b}_2}{dt} = -i(2\omega_2)\hat{b}_2 - \frac{g}{2}\hat{a}_2^{\,2} \tag{23}$$

These equations characterize the temporal evolution of the mode annihilation operator in the system. These are coupled differential equations that cannot be solved in a closed analytical form. As such, this method aims to obtain its closed-form solution by employing the weak-nonlinearity approximation, where the nonlinearity of the system is restricted to its linear term only. In another word, the terms g^2 and above are prohibited from the system solution. According to the Heisenberg picture of quantum mechanics, the solution to Eqs. (20)–(23) can be expressed as

$$\hat{O}_j = \exp\left(\frac{i\hat{H}t}{\hbar}\right)\hat{O}_j(0)\exp\left(-\frac{i\hat{H}t}{\hbar}\right) \tag{24}$$

In fact, this equation resembles the Baker-Campbell-Hausdorff formula, and it can also be expanded in the Taylor series to yield

$$\hat{O}_j = \hat{O}_j(0) + \frac{it}{\hbar}\left[\hat{H},\hat{O}_j(0)\right] - \frac{1}{2!}\left(\frac{t^2}{\hbar^2}\right)\left[\hat{H},\left[\hat{H},\hat{O}_j(0)\right]\right] + \ldots \tag{25}$$

The commutators $[\hat{H},\hat{O}_j(0)]$ and $[\hat{H},[\hat{H},\hat{O}_j(0)]\,]$ can be evaluated using the initial state of Eqs. (20)–(23). Upon the evaluation of these commutators, the solution can be assumed to have the following form:

$$\begin{aligned}\hat{a}_1(t) = {} & A_1\hat{a}_1(0) + A_2\hat{a}_2(0) + A_3\hat{a}_1^{\dagger}(0)\hat{b}_1(0) \\ & + A_4\hat{a}_2^{\dagger}(0)\hat{b}_1(0) + A_5\hat{a}_2^{\dagger}(0)\hat{b}_2(0) \\ & + A_6\hat{a}_1(0)\hat{b}_1^{\dagger}(0)\hat{b}_1(0) + A_7\hat{a}_1^{\dagger}(0)\hat{a}_1^{\,2}(0)\end{aligned} \tag{26}$$

$$\hat{a}_2(t) = B_1\hat{a}_1(0) + B_2\hat{a}_2(0) + B_3\hat{a}_2^{\dagger}(0)\hat{b}_2(0)$$
$$+B_4\hat{a}_1^{\dagger}(0)\hat{b}_2(0) + B_5\hat{a}_1^{\dagger}(0)\hat{b}_1(0)$$
$$+B_6\hat{a}_2^{\dagger}(0)\hat{a}_2^{2}(0) + B_7\hat{a}_2(0)\hat{b}_2^{\dagger}(0)\hat{b}_2(0) \quad (27)$$

$$\hat{b}_1(t) = \bar{A}_1\hat{b}_1(0) + \bar{A}_2\hat{a}_1^{2}(0) + \bar{A}_3\hat{a}_2(0)\hat{a}_1(0)$$
$$+\bar{A}_4\hat{a}_1^{\dagger}(0)\hat{a}_1(0)\hat{b}_1(0) + \bar{A}_5\hat{a}_1(0)\hat{a}_1^{\dagger}(0)\hat{b}_1(0) \quad (28)$$

$$\hat{b}_2(t) = \bar{B}_1\hat{b}_2(0) + \bar{B}_2\hat{a}_2^{2}(0) + \bar{B}_3\hat{a}_1(0)\hat{a}_2(0)$$
$$+\bar{B}_4\hat{a}_2^{\dagger}(0)\hat{a}_2(0)\hat{b}_2(0) + \bar{B}_5\hat{a}_2(0)\hat{a}_2^{\dagger}(0)\hat{b}_2(0) \quad (29)$$

where $\{A_m\}_{m=1}^{7}$, $\{B_m\}_{m=1}^{7}$, $\{\bar{A}_m\}_{m=1}^{5}$ and $\{\bar{B}_m\}_{m=1}^{5}$ are time-dependent functions. In the attempt to obtain these functions, the solutions in Eqs. (26)–(29) are substituted into the Heisenberg equation of motion Eqs. (20) (23), and equating the common terms from both the sides to get the following set of system:

$$\frac{dA_1}{dt} = -i\omega_1 A_1 - i\kappa B_1 \quad (30)$$

$$\frac{dA_2}{dt} = -i\omega_1 A_2 - i\kappa B_2 \quad (31)$$

$$\frac{dA_3}{dt} = -i\omega_1 A_3 + gA_1^{*}\bar{A}_1 - i\kappa B_5 \quad (32)$$

$$\frac{dA_4}{dt} = -i\omega_1 A_4 + gA_2^{*}\bar{A}_1 \quad (33)$$

$$\frac{dA_5}{dt} = -i\omega_1 A_5 - i\kappa B_3 \quad (34)$$

$$\frac{dA_6}{dt} = -i\omega_1 A_6 + gA_3^{*}\bar{A}_1 \quad (35)$$

$$\frac{dA_7}{dt} = -i\omega_1 A_7 + gA_1^* \bar{A}_2 \tag{36}$$

Similarly, the following differential systems for coefficients B_m $\bar{A}_m$ and $\bar{B}_m$ can be expressed as:

$$\frac{dB_1}{dt} = -i\omega_2 B_1 - i\kappa A_1 \tag{37}$$

$$\frac{dB_2}{dt} = -i\omega_2 B_2 - i\kappa A_2 \tag{38}$$

$$\frac{dB_3}{dt} = -i\omega_2 B_3 + gB_2^* \bar{B}_1 - i\kappa A_5 \tag{39}$$

$$\frac{dB_4}{dt} = -i\omega_2 B_4 + gB_1^* \bar{B}_1 \tag{40}$$

$$\frac{dB_5}{dt} = -i\omega_2 B_5 - i\kappa A_3 \tag{41}$$

$$\frac{dB_6}{dt} = -i\omega_2 B_6 + gB_2^* \bar{B}_2 \tag{42}$$

$$\frac{dB_7}{dt} = -i\omega_2 B_7 + gB_3^* \bar{B}_1 \tag{43}$$

$$\frac{d\bar{A}_1}{dt} = -i(2\omega_1)\bar{A}_1 \tag{44}$$

$$\frac{d\bar{A}_2}{dt} = -i(2\omega_1)\bar{A}_2 - \frac{g}{2}A_1^{\,2} \tag{45}$$

$$\frac{d\bar{A}_3}{dt} = -i(2\omega_1)\bar{A}_3 - gA_1A_2 \tag{46}$$

$$\frac{d\bar{A}_4}{dt} = -i(2\omega_1)\bar{A}_4 - gA_3A_1 \quad (47)$$

$$\frac{d\bar{A}_5}{dt} = -i(2\omega_1)\bar{A}_5 - gA_1A_3 \quad (48)$$

$$\frac{d\bar{B}_1}{dt} = -i(2\omega_2)\bar{B}_1 \quad (49)$$

$$\frac{d\bar{B}_2}{dt} = -i(2\omega_2)\bar{B}_2 - \frac{g}{2}B_2^{\,2} \quad (50)$$

$$\frac{d\bar{B}_3}{dt} = -i(2\omega_2)\bar{B}_3 - gB_1B_2 \quad (51)$$

$$\frac{d\bar{B}_4}{dt} = -i(2\omega_2)\bar{B}_4 - \frac{g}{2}B_3B_2 \quad (52)$$

$$\frac{d\bar{B}_5}{dt} = -i(2\omega_2)\bar{B}_5 - \frac{g}{2}B_2B_3 \quad (53)$$

This set of differential equations is then solved numerically using the fourth-order Runge Kutta method.

2.3. Scaling of The System

For the convenience of numerical simulation, the system is scaled into the dimensionless form by expressing all the parameters in the ω_1 basis (for a codirectional system). For the phase space method, the scaling is performed on the stochastic differential Eqs. (11)–(18), where both sides of these equations are divided with the ω_1 to yield

$$\frac{d\alpha_1}{d\tau} = -i\tilde{\omega}_1\alpha_1 + \tilde{g}\bar{\alpha}_1\beta_1 - i\tilde{\kappa}\alpha_2 + \sqrt{\tilde{g}\bar{\alpha}_1}\eta_1(\tau) \quad (54)$$

$$\frac{d\beta_1}{d\tau}=i\tilde{\omega}_1\beta_1+\tilde{g}\alpha_1\bar{\beta}_1+i\tilde{\kappa}\beta_2+\sqrt{\tilde{g}\bar{\beta}_1}\eta_2(\tau) \tag{55}$$

$$\frac{d\alpha_2}{d\tau}=-i\tilde{\omega}_2\alpha_2+\tilde{g}\bar{\alpha}_2\beta_2-i\tilde{\kappa}\alpha_1+\sqrt{\tilde{g}\bar{\alpha}_2}\eta_3(\tau) \tag{56}$$

$$\frac{d\beta_2}{d\tau}=i\tilde{\omega}_2\beta_2+\tilde{g}\alpha_2\bar{\beta}_2+i\tilde{\kappa}\beta_1+\sqrt{\tilde{g}\bar{\beta}_2}\eta_4(\tau) \tag{57}$$

$$\frac{d\alpha_3}{d\tau}=-i(2\tilde{\omega}_1)\bar{\alpha}_1-\frac{\tilde{g}}{2}\alpha_1^{\,2} \tag{58}$$

$$\frac{d\bar{\beta}_1}{d\tau}=i(2\tilde{\omega}_1)\bar{\beta}_1-\frac{\tilde{g}}{2}\beta_1^{\,2} \tag{59}$$

$$\frac{d\bar{\alpha}_2}{d\tau}=-i(2\tilde{\omega}_2)\bar{\alpha}_2-\frac{\tilde{g}}{2}\alpha_2^{\,2} \tag{60}$$

$$\frac{d\bar{\beta}_2}{d\tau}=i(2\tilde{\omega}_2)\bar{\beta}_2-\frac{\tilde{g}}{2}\beta_2^{\,2} \tag{61}$$

where $\widetilde{\omega}_1 = \frac{\omega_1}{\omega_1} = 1$, $\widetilde{\omega}_2 = \frac{\omega_2}{\omega_1}$, $\tilde{g} = \frac{g}{\omega_1}$, $\tilde{\kappa} = \frac{\kappa}{\omega_1}$, $\tau = \omega_1 t$ and $\tilde{\eta}(\tau) = \frac{\tilde{\eta}(t)}{\sqrt{\omega_1}}$ are the dimensionless parameters. For the analytical method, the scaling is performed on Eqs. (30)–(53). As a result, all the parameters that have a dimension is replaced by its dimensionless counterparts, i.e., $\omega_1 \rightarrow \widetilde{\omega}_1$, $\omega_2 \rightarrow \widetilde{\omega}_2$, $g \rightarrow \tilde{g}$, $\kappa \rightarrow \tilde{\kappa}$ and $t \rightarrow \tau$.

2.4. Squeezing in Codirectional Two-Channel System

Squeezed states are those at which the uncertainty distribution in one of the quadrature components is reduced below the symmetric limit of a coherent state at the expense of the increased uncertainty in the other quadrature, such that the Heisenberg uncertainty relation is not violated [34]. In other words, in a squeezed state, the circle denoting the uncertainty of a coherent state (see Figure 2(a)) in the quadrature-phase space has been "squeezed" to an ellipse of the same area (see Figure 2(b)–(c)). That means, one component of quadrature variances has a value below the standard quantum limit.

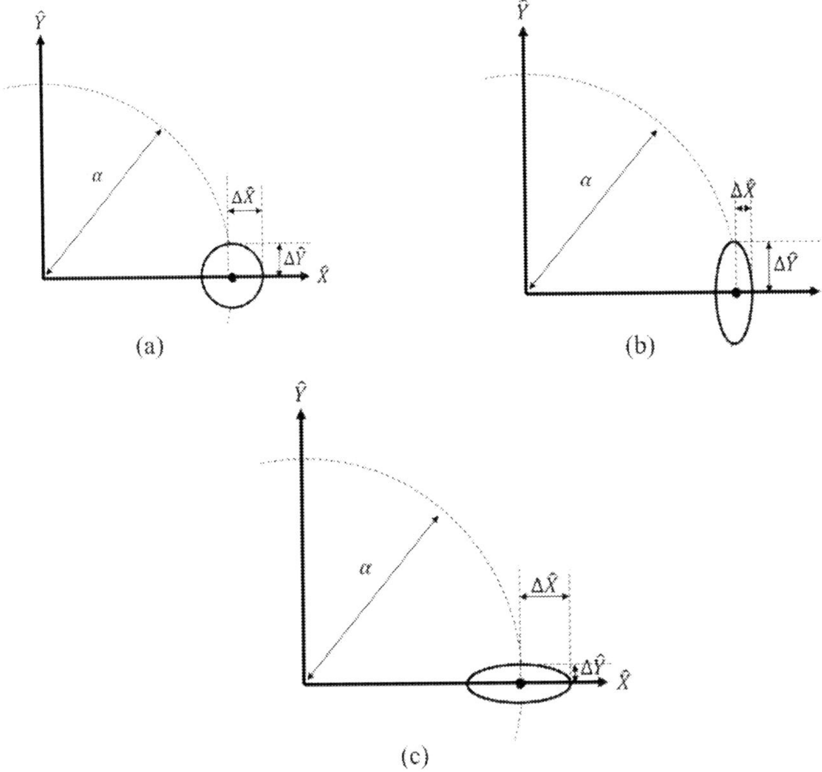

Figure 2. Uncertainty distribution for (a) coherent state, and squeezed uncertainty distribution in the (b) first, and (c) second quadrature components.

$$\left\langle (\Delta\hat{X})^2 \right\rangle < \frac{1}{4} \text{ or } \left\langle (\Delta\hat{Y})^2 \right\rangle < \frac{1}{4} \tag{62}$$

where $\langle(\Delta\hat{X})^2\rangle$ and $\langle(\Delta\hat{Y})^2\rangle$ are the first and the second quadrature variances. However, Eq. (62) must agree with the Heisenberg's uncertainty relation, where:

$$\sqrt{\left\langle (\Delta\hat{X})^2 \right\rangle \left\langle (\Delta\hat{Y})^2 \right\rangle} \geq \frac{1}{4} \tag{63}$$

In the former inequality, $\sqrt{\langle(\Delta\hat{X})^2\rangle}$ and $\sqrt{\langle(\Delta\hat{Y})^2\rangle}$ are, respectively, associated with the uncertainty in position Δr and momentum Δp. For better visualization, squeezed states can be illustrated in Figure 2(b, c). The uncertainty distribution for squeezed states assumes the form of an ellipse, instead of a circle for the coherent states. Figure 2(b) shows the uncertainty reduction in the first quadrature component, while Figure 2(c) shows the uncertainty reduction in the second quadrature component.

From the above explanation, squeezing can be identified by evaluating the quadrature variances for both the quadrature components. Further detail on this will be explained in the following section.

2.5. Squeezing Measurement

Squeezing can mainly be investigated by calculating the variances of the quadrature operator. Quadrature operators for the first and second quadrature components are given as:

$$\hat{X}_j = \frac{1}{2}\left(\hat{O}_j + \hat{O}_j^{\dagger}\right) \tag{64}$$

$$\hat{Y}_j = -\frac{i}{2}\left(\hat{O}_j - \hat{O}_j^{\dagger}\right) \tag{65}$$

and its variances can be evaluated through the following expressions:

$$\left\langle (\Delta\hat{X})^2 \right\rangle = \langle \hat{X}^2 \rangle - \langle \hat{X} \rangle^2 \tag{66}$$

$$\left\langle (\Delta\hat{Y})^2 \right\rangle = \langle \hat{Y}^2 \rangle - \langle \hat{Y} \rangle^2 \tag{67}$$

where $\hat{X}$ and $\hat{Y}$ are the quadrature operators for the first and second quadrature components, respectively. By substituting Eqs. (64), (65) into Eqs. (66), (67), the following equation can be derived:

$$\left\langle (\Delta\hat{X})^2 \right\rangle, \left\langle (\Delta\hat{Y})^2 \right\rangle = \frac{1}{4}\Big[1 + 2\langle \hat{O}^{\dagger}\hat{O} \rangle - 2\langle \hat{O} \rangle \langle \hat{O}^{\dagger} \rangle \pm \left(\langle \hat{O}^2 \rangle + \langle \hat{O}^{\dagger 2} \rangle - \langle \hat{O} \rangle^2 - \langle \hat{O}^{\dagger} \rangle^2 \right)\Big] \tag{68}$$

These equations are the quadrature variances expressed in terms of mode annihilation and creation operators. Therefore, it is more convenient to use this equation for further derivation, since the solution obtained here is mainly in the form of a mode-annihilation operator.

Exceptionally, in the phase space method, $\hat{a}$ and $\hat{a}^{\dagger}$ are expressed in the classical form, where $\hat{a} \rightarrow \alpha$ and $\hat{a}^{\dagger} \rightarrow \beta$. Therefore, Eq. (68) needs to be modified to become:

$$\left\langle (\Delta\hat{X})^2 \right\rangle, \left\langle (\Delta\hat{Y})^2 \right\rangle = \frac{1}{4}\Big[1 + 2\langle \beta\alpha \rangle - 2\langle \alpha \rangle \langle \beta \rangle \pm \left(\langle \alpha^2 \rangle + \langle \beta^2 \rangle - \langle \alpha \rangle^2 - \langle \beta \rangle^2 \right)\Big] \tag{69}$$

Using this method, the entire calculation is performed numerically. Therefore, the final form of quadrature variance cannot be expressed in the equation form.

For the analytical method, squeezing in the first mode can be obtained by substituting the solution Eq. (26) into Eq. (68). Meanwhile, squeezing in the second mode can be identified by substituting the solution Eq. (27) into the expressions in Eq. (68). The operation yields:

$$\left\langle (\Delta\hat{X}_{a_1})^2 \right\rangle, \left\langle (\Delta\hat{Y}_{a_1})^2 \right\rangle = \frac{1}{4}\Big[1 + 2\left(|A_5|^2 |\beta_2|^2 + A_5 A_4^* \beta_1^* \beta_2 + A_4 A_5^* \beta_2^* \beta_1 \right) \pm \left(A_3 A_1 \beta_1 + A_7 A_1 \alpha_1^2 + A_4 A_2 \beta_1 + A_5 A_2 \beta_2 + \text{c.c.} \right)\Big] \tag{70}$$

$$\left\langle (\Delta\hat{X}_{a_2})^2 \right\rangle, \left\langle (\Delta\hat{Y}_{a_2})^2 \right\rangle = \frac{1}{4}\Big[1 + 2\left(|B_5|^2 |\beta_1|^2 + B_5 B_4^* \beta_2^* \beta_1 + B_4 B_5^* \beta_1^* \beta_2 \right) \pm \left(B_4 B_1 \beta_2 + B_5 B_1 \beta_1 + B_3 B_2 \beta_2 + B_6 B_2 \alpha_2^2 + \text{c.c.} \right)\Big] \tag{71}$$

where c.c. is the complex conjugate and α_j that corresponds to the initial mode-annihilation operator $\hat{a}_j(0)$. From the quadrature variances, squeezing can be indicated by the region that is below the shot noise level, i.e., $\langle (\Delta\hat{X})^2 \rangle < \frac{1}{4}$ or $\langle (\Delta\hat{Y})^2 \rangle < \frac{1}{4}$.

3. Results and Discussion

In this section, the graphical results obtained from the previous solution are presented. Here, the variation of quadrature variances is plotted against the dimensionless time (τ). For all the graphs, the system is initialized symmetrically using the coherent light with the amplitude $\alpha_1 = \alpha_2 = 1$. In Figure 3, squeezing generated in the first (Figure 3(a)) and second (Figure 3(b)) modes are presented. These results show that both the modes produced squeezed light with a steady oscillation pattern, and this behaviour is recorded by both the analytical and phase space method. High agreement between both the methods is achieved at the early evolution stage, but this agreement becomes lower starting at τ~4.

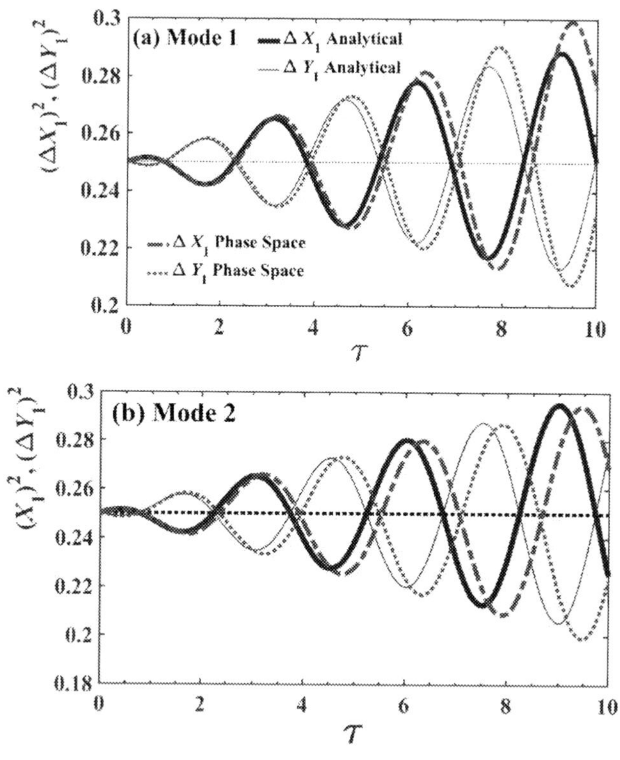

Figure 3. Field Quadrature variances for (a) mode $\hat{a}_1$, and (b) mode $\hat{a}_2$ for the input parameters $\alpha_1 = \alpha_2 = 1$, $\widetilde{\omega}_1 = \widetilde{\omega}_2 = 1$, $\tilde{g} = 0.01$ and $\tilde{\kappa} = 0.1$. Here and in all subsequent graphs the horizontal line represents the shot-noise level.

Here, the effect of linear coupling is identified by setting different values of linear coupling constant. This parameter can be tuned by changing the separation distance between the channels, where large separation effect in weak linear coupling (or low $\tilde{\kappa}$). We choose $\tilde{\kappa} = 0.01$, $\tilde{\kappa} = 0.09$ and $\tilde{\kappa} = 0.5$ (see Figure 4(a)–(c), respectively). When $\tilde{\kappa} = 0.01$, both the methods agree with each other almost every single time. When $\tilde{\kappa} = 0.09$, the disagreement appears starting at $\tau \sim 4$, but it is not significant. When $\tilde{\kappa} = 0.5$, the phase space method significantly disagrees with the analytical method, where the agreement is maintained only up to $\tau \sim 1$. Therefore, as linear coupling interaction becomes stronger, the agreement between the phase space and analytical methods decreases. Thus, these two methods can be said to be complementary to each other (rather than an alternative to each other).

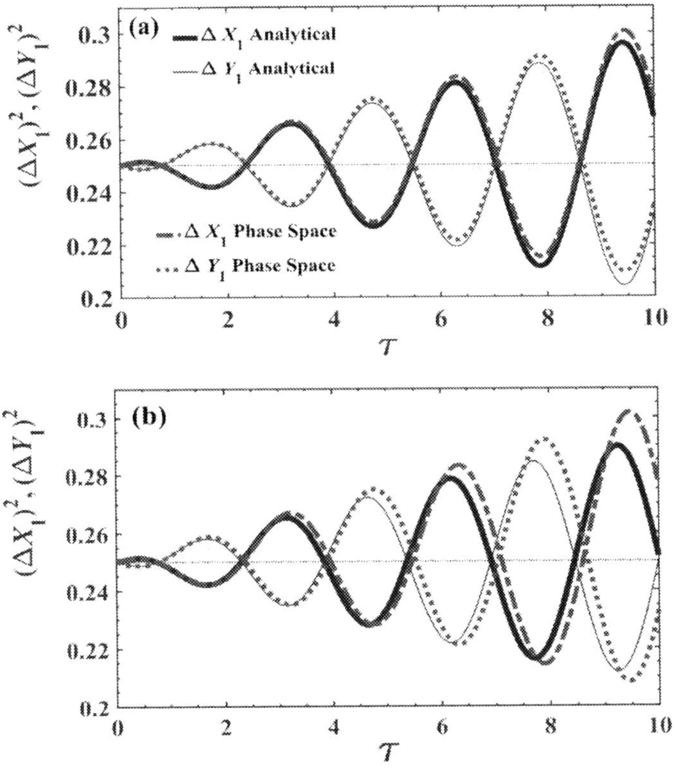

Figure 4. Temporal evolution of quadrature variances for the mode $\hat{a}_1$ at different linear couplings; (a) $\tilde{\kappa} = 0.01$, (b) $\tilde{\kappa} = 0.09$, and (c) $\tilde{\kappa} = 0.5$. Other parameters are set similar to Figure 3.

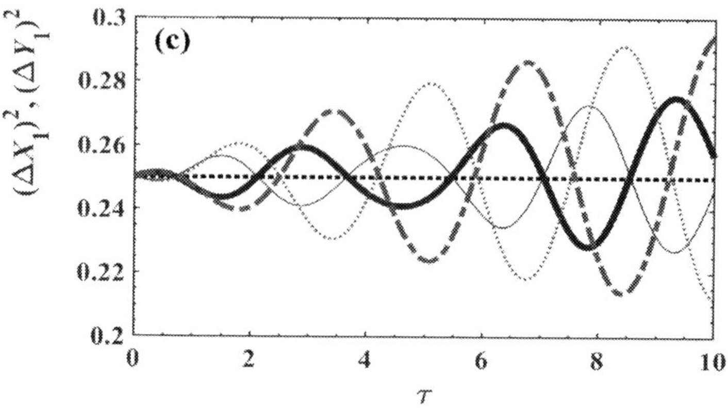

Figure 4. (Continued)

In Figure 5, we present the effect of nonlinear coupling on the generated squeezed light. We choose, $\tilde{g} = 0.05$ (Figure 5(a)) and $\tilde{g} = 0.08$ (Figure 5(b)) to study this effect. For both the results, the phase space agrees with the analytical method at the early evolution stage (up to $\tau \sim 2$) only. When the nonlinear coupling constant increases, we observe the improvement in maximal squeezing amplitude, i.e., the squeezing signal is stronger at a high value of $\tilde{g}$. In addition, when $\tilde{g}$ increases, the agreement between both the methods becomes lower.

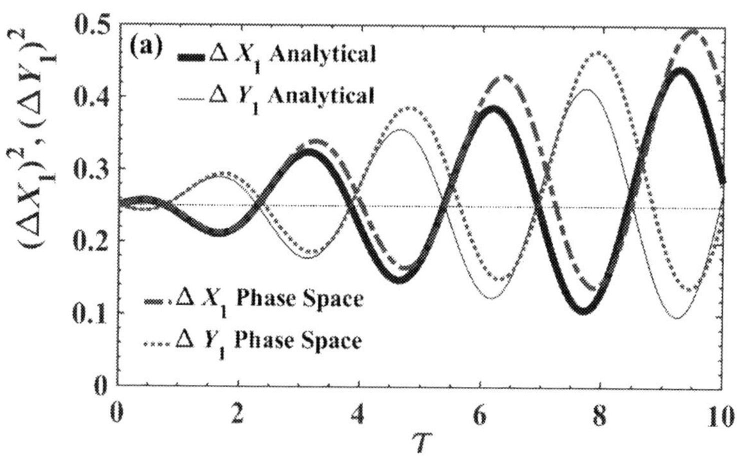

Figure 5. Fluctuation of squeezing in the mode $\hat{a}_1$ at different nonlinear couplings; (a) $\tilde{g} = 0.05$ and (b) $\tilde{g} = 0.08$. Here, the linear coupling constant is set as $\tilde{\kappa} = 0.1$, and other parameters are $\alpha_1 = \alpha_2 = 1$, $\tilde{\omega}_1 = \tilde{\omega}_2 = 1$.

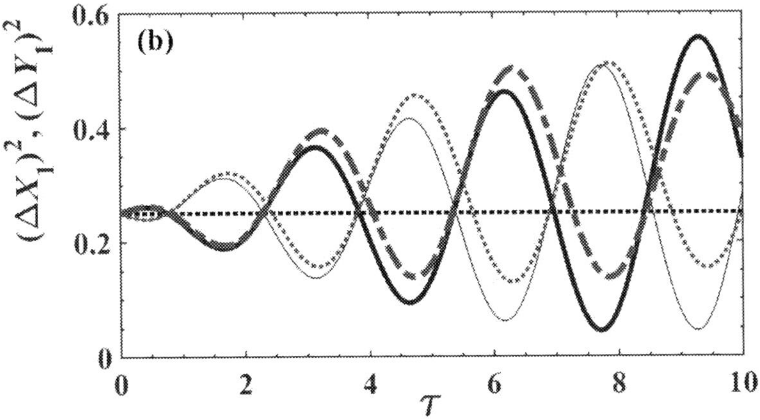

Figure 5. (Continued)

In Figure 6, the variation of quadrature variances in the first mode is plotted under the frequency mismatch condition. Under this condition, the frequency of each fundamental mode is different (i.e., $\Delta\tilde{\omega} = |\tilde{\omega}_2 - \tilde{\omega}_1| \neq 0$). In obtaining these results, we set $\alpha_1 = \alpha_2 = 1$, $\tilde{g} = 0.01$ and $\tilde{\kappa} = 0.2$. The effect of frequency mismatch is investigated by plotting a graph for different values of frequency mismatch. Specifically, $\Delta\tilde{\omega} = 0.5$, $\Delta\tilde{\omega} = 1$ and $\Delta\tilde{\omega} = 8$ are chosen (refer to Figure 6(a)–(c), respectively). From these results, we can see that the phase space method yields a similar graph for all values of $\Delta\tilde{\omega}$. Meanwhile, for the analytical method, the effect of frequency mismatch is significant, where different oscillation patterns are generated for each value of $\Delta\tilde{\omega}$. Also, the analytical method illustrated the increase in maximal squeezing amplitude when the frequency mismatch increases. From this observation, we can conclude that the phase space method is less sensitive upon detecting the effect of frequency mismatch in the first mode. In addition, a high agreement between the phase space and analytical methods is achieved at high $\Delta\tilde{\omega}$.

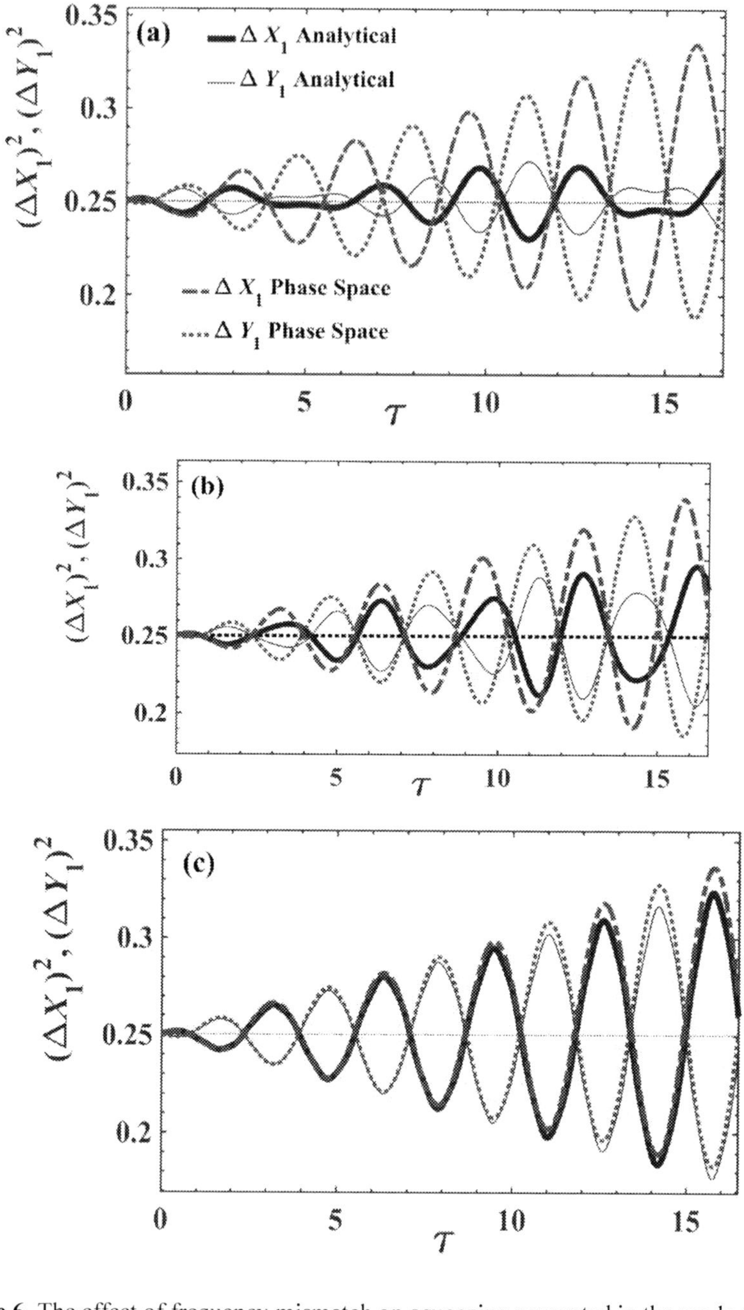

Figure 6. The effect of frequency mismatch on squeezing generated in the mode $\hat{a}_1$ for (a) $\Delta\tilde{\omega} = 0.5$, (b) $\Delta\tilde{\omega} = 1$, and (c) $\Delta\tilde{\omega} = 8$.

The effect of frequency mismatch on the second mode is presented in Figure 7. Here we assign $\Delta\widetilde{\omega} = 1$ in Figure 7(a), $\Delta\widetilde{\omega} = 4$ in Figure 7(b); the other parameters are set similar to those in Figure 6. Here, we obtained the evolution of squeezing with stronger oscillation compared to Figure 6. It is because the frequency of the second mode is higher compared to the first mode. Also, the fluctuation is more violent in Figure 7(b) (with $\widetilde{\omega}_2 = 5$) compared to Figure 7(a) (with $\widetilde{\omega}_2 = 2$), and this proved that the squeezing frequency is highly dependent on the frequency of the input field. From these results, the analytical method produces a collapsed-revival pattern for both the values of $\Delta\widetilde{\omega}$. On the other side, the phase space method generates a squeezing with a steady oscillation pattern. This explains that the phase space method is less sensitive to the effect of frequency mismatch under this combination of input parameters.

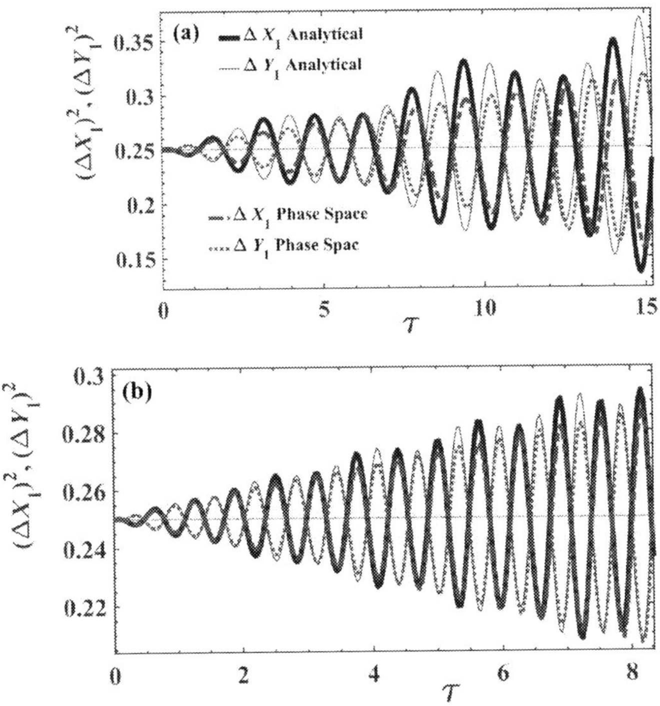

Figure 7. The effect of frequency mismatch on squeezing generated in the mode $\hat{a}_2$ for (a) $\Delta\widetilde{\omega} = 1$ and (b) $\Delta\widetilde{\omega} = 4$.

Conclusion

In conclusion, the two-channel coupler with SHG has been formulated for the investigation of squeezing. The evolution of the mode operator has been solved using the phase space and analytical methods. Under the frequency matching, both the methods generated a steady oscillation pattern, and the agreement between both the methods is achieved mostly at the early propagation period. However, at weak linear coupling interaction, the agreement is continued for a long period (refer to Figure 4(a)), and this agreement becomes faded as the linear coupling increases. Under frequency mismatch, the analytical perturbative method generates a mild collapsed-revival-like pattern in the first and the second modes. Inversely, the phase space method maintains its steady oscillation pattern for all values of $\Delta\tilde{\omega}$. These outcomes explain the insensitivity of the phase space method to sense the existence of frequency mismatch for the chosen input parameters.

Acknowledgments

Abdel-Baset M. A. Ibrahim gratefully recognizes the Malaysian Ministry of Higher Education (MOHE) national grant FRGS/1/2021/STG07/UITM/02/10 and the internal grant 600-RMC/MYRA 5/3/LESTARI (069/2020) from Universiti Teknologi MARA (UiTM).

References

[1] Tameshtit, "Coherent processing of a qubit using one squeezed state," *Entropy*, vol. 19, no. 12, p. 653, 2017.

[2] Lie S. H. and H. Jeong, "Limitations of teleporting a qubit via a two-mode squeezed state" *Photonics Res.*, vol. 7, no. 5, pp. A7-A13, 2019.

[3] Ralph T. C. and P. K. Lam, "Teleportation with bright squeezed light," *Phys. Rev. Lett.*, vol. 81, no. 25, pp. 5668–5671, 1998.

[4] Anderson B. E. et al. "Phase sensing beyond the standard quantum limit with a variation on the SU(1,1) interferometer," *Optica*, vol. 4, no. 7, p. 752, 2017.

[5] Mansell G. L., "*Squeezed Light Sources for Current and Future Interferometric Gravitational-Wave Detectors*," *Ph. D thesis*, Australian National University, 2018.

[6] Sohma M., "Transmission of classical information with squeezed states" *Tamagawa University Quantum ICT Research Bulletin.*, vol. 3, no. 1, pp. 13-20, 2013.

[7] Matthews J. C. F., A. Politi, A. Stefanov, and J. L. O'Brien, "Manipulating multi-photon entanglement in waveguide quantum circuits," *Nat. Photonics*, vol. 3, no. 6, pp. 346–350, 2009.

[8] Politi A., J. C. F. Matthews, and J. L. O'Brien, "Shor's quantum factoring algorithm on a photonic chip," *Science*, vol. 325, no. 5945, pp. 1221–1221, 2009.

[9] Mandal P. and S. Midda, "All optical method of developing OR and NAND logic system based on nonlinear optical fiber couplers," *Optik*, vol. 122, no. 20, pp. 1795–1798, 2011.

[10] Mukherjee K. and P. C. Jana, "Higher-order squeezing of a generic quadratically-coupled optomechanical system," *IOSR J. Appl. Phys.*, vol. 9, no. 01, pp. 90–95, 2017.

[11] Mukherjee K. and P. C. Jana, "Squeezing and entanglement in quadratically-coupled optomechanical system," *J. Phys. Sci.*, vol. 19, pp. 143-155, 2014.

[12] Sen B., V. Peřinová, A. Lukš, J. Peřina, and J. Křepelka, "Sub-shot noise photon-number correlation in stimulated and spontaneous Raman processes," *J. Phys. B At. Mol. Opt. Phys.*, vol. 44, no. 10, p. 105503, 2011.

[13] Sen B. and S. Mandal, "Squeezed states in spontaneous Raman and in stimulated Raman processes," *J. Mod. Opt.*, vol. 52, no. 13, pp. 1789–1807, 2005.

[14] Giri S. K., B. Sen, A. Pathak, and P. C. Jana, "Higher-order two-mode and multimode entanglement in Raman processes," *Phys. Rev. A*, vol. 93, no. 1, p. 012340, 2016.

[15] Thapliyal K., A. Pathak, B. Sen, and J. Perina, "Nonclassicality in non-degenerate hyper-Raman processes," *Quant-Ph*, 2017.

[16] Kumar G. S., S. Biswajit, and J. P. Chandra, "General higher order intermodal antibunching in two-mode Bose Einstein condensates," *Res. J. Recent Sci.*, vol. 5, pp. 13-16, 2016.

[17] Mukhopadhyay A., B. Sen, K. Thapliyal, S. Mandal, and A. Pathak, "Interaction of light and semiconductor can generate quantum states required for solid-state quantum computing: entangled, steered and other nonclassical states," *Quantum Inf. Process.*, vol. 18, no. 7, p. 234, 2019.

[18] Hanapi M. S. M., A.-B. M. A. Ibrahim, R. Julius, and H. Eleuch, "The quantum Kerr nonlinear coupler: the analytical versus phase-space method," *Can. J. Phys.*, vol. 99, no. 9, 2021.

[19] Hanapi M. S. M., A.-B. M. A. Ibrahim, and P. K. Choudhury, "On the perturbative approach to analyse entanglement in two-channel Kerr nonlinear coupler," *Optik*, vol. 243, p. 167420, 2021.

[20] Thapliyal K., A. Pathak, B. Sen, and J. Perina, "Higher order nonclassicalities in a codirectional nonlinear optical coupler: Quantum entanglement, squeezing and antibunching," *Phys. Rev. A*, vol. 90, no. 1, p. 013808, 2014.

[21] Thapliyal K. and A. Pathak, "Quantum Zeno and anti-Zeno effects in an asymmetric nonlinear optical coupler," In *International Conference on Optics and Photonics 2015*, vol. 9654, pp. 282-288. SPIE, 2015.

[22] Mandal S. and J. Perina, "Approximate quantum statistical properties of a nonlinear optical coupler," *Phys. Lett. A*, vol. 328, no. 2–3, pp. 144–156, 2004.

[23] Thapliyal K., A. Pathak, B. Sen, and J. Perina, "Nonclassical properties of a contradirectional nonlinear optical coupler," *Phys. Lett. A*, vol. 378, no. 46, pp. 3431–3440, 2014.

[24] Ibrahim A.-B. M. A., B. A. Umarov, and M. R. B. Wahiddin, "Squeezing in the Kerr nonlinear coupler via phase-space representation," *Phys. Rev. A*, vol. 61, no. 4, p. 043804, 2000.

[25] Umarov B. A., A.-B. M. A. Ibrahim, M. R. B. Wahiddin, Quantum Dynamics in Kerr Nonlinear Coupler via Phase-Space Representation. In "*Nonlinearity and Disorder: Theory and Applications,*" Ed., F. Abdullaev, O. Bang, M.P. Sørensen, NATO Science Series (Series II: Mathematics, Physics and Chemistry), vol. 45, pp. 341-349 Springer, Dordrecht (2001).

[26] Julius R., A.-B. M. A. Ibrahim, and M. S. M. Deni, "Quantum state generation in a four-mode Kerr nonlinear directional coupler," *Laser Phys.*, vol. 24, no. 3, p. 035202, 2014.

[27] Julius R. and A.-B. M. A. Ibrahim, "Quantum squeezing in multichannel nonlinear coupler," *JCSCM*, pp. 131–136, 2017.

[28] Julius R., A.-B. M. A. Ibrahim, M. K. Abd-Rahman, et al. Quantum dynamics of a four-channel Kerr nonlinear directional coupler. *Opt. Rev.*, vol. 25, pp. 563–570, 2018.

[29] Julius R., A.-B. M. A. Ibrahim, P. K. Choudhury, and H. Eleuch, "On the nonclassical dynamics of cavity-assisted four-channel nonlinear coupler," *Chinese Phys. B*, vol. 27, no. 11, p. 114206, 2018.

[30] Julius R., A.-B. M. A. Ibrahim and Pankaj K. Choudhury, Dielectric Waveguides as a Source for Nonclassical Light. In "*Dielectric Materials and Applications,*" Ed., P. K. Choudhury, Nova science publishers, USA, pp. 335–358 (2019).

[31] Julius R., A.-B. M. A. Ibrahim, P. K. Choudhury, and H. Eleuch, "Quantum states generation in multichannel directional coupler with second-order nonlinearity," *Optik*, vol. 186, pp. 212–220, 2019.

[32] Julius R., A.-B. M. A. Ibrahim, H. Eleuch, and P. K. Choudhury, "Sub-Poissonian photon squeezing and entanglement in optical chain second harmonic generation," *J. Mod. Opt.* vol. 66, no. 10, pp. 1129–1138, 2019.

[33] Julius R., M. Zahirzai, A.-B. M. A. Ibrahim, H. Eleuch, and P. K. Choudhury, "Single- and compound-mode squeezing in nonlinear coupler with frequency mismatch," *J. Electromagn. Waves Appl.*, vol. 34, no. 3, pp. 301–315, 2020.

[34] Ficek Z. and M. R. Wahiddin, *Quantum Optics for Beginners*, 1st ed. Jenny Stanford Publishing (2016).

Chapter 7

Quantum Dynamics of Contradirectional Nonlinear Coupler

Mohd Syafiq M. Hanapi[1], **Abdel-Baset M. A. Ibrahim**[1,*], **Rafael Julius**[2], **Ahmad M. Alrub**[3] **and Pankaj K. Choudhury**[4]

[1]Faculty of Applied Sciences, Universiti Teknologi MARA (UiTM), Shah Alam, Selangor, Malaysia
[2]Faculty of Applied Sciences, Universiti Teknologi MARA (UiTM) Perak, Tapah Campus, Perak, Malaysia
[3]Department of Physics, Al-Hussein Bin Talal University, Ma'an 71111, Jordan
[4]Institute of Microengineering and Nanoelectronics, Universiti Kebangsaan Malaysia, UKM Bangi, Selangor, Malaysia

Abstract

This chapter presents the research results of studies on squeezing in coupler system with contra-directional propagation. The second harmonic generation in this system is driven by the second-order nonlinearity of waveguides.

The investigation is performed using the phase space and analytical methods; the comparison between them remains the main focus in this chapter.

Keywords: Quantum optics, second harmonic generation, optical coupler, squeezed states

[*] Corresponding Author's Email: abdelbaset@uitm.edu.my.

In: Second Harmonic Generation
Editors: Abdel-Baset M. A. Ibrahim and Pankaj Kumar Choudhury
ISBN: 978-1-68507-888-1

1. Introduction

Quantum states of light hold great promise in a wide range of important applications. Squeezed states of light, for example, can be employed in precision measurements [1, 2], binary quantum communication [3, 4], sensing, and tracking [5]. Squeezed states of light could also be used to empower the futuristic quantum computer. However, the viability of squeezed light in quantum computing depends on its ability to initiate effective quantum teleportation [6-8]. Because of its intriguing potential, studies in this area have recently become one of the main attractions. Researchers are still looking for reliable quantum light producers [9-11] as well as the best mathematical techniques that could be applied [12, 13]. Experiments have also been performed to detect these quantum states in the laboratory [14, 15].

In this chapter, we will investigate the squeezed states of light generated in two-channel coupler with second-order nonlinearity in the case of contra-directional propagation. We analyze the scenario of second-harmonic generation (SHG), in which photons from the pump field (fundamental photons) are up-converted in pairs to double-frequency second harmonic photons. Figure 1 depicts the schematic diagram of the system under consideration. The following discussion is based on theoretical results gained through the use of phase space and analytical methods.

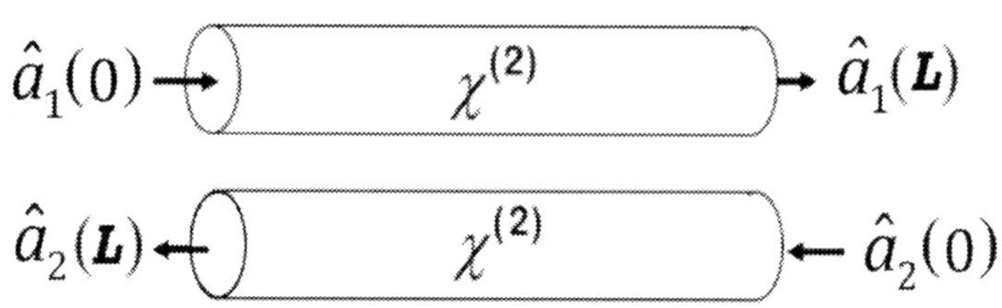

Figure 1. Schematic diagram of the contra-directional two-channel symmetric coupler with second-order nonlinearity.

The phase space method is based on the Schrödinger equation, where the quantum operator is time-independent, while the state vector is time-dependent. Hence, the evolution of the system in the phase space method will be specified by the state vector, not the operator. The time-evolution of a state vector is governed by the following Schrödinger equation:

$$\frac{\partial}{\partial t}|\psi_S(t)\rangle = -\frac{i}{\hbar}\hat{H}|\psi_S(t)\rangle \tag{1}$$

In the above equation, the subscript S indicates the Schrödinger picture; hence, ψ_S represents the state vector in the Schrödinger picture. The above equation has the following form of solution [16]:

$$|\psi_S(t)\rangle = \hat{U}(t)|\psi_S(0)\rangle \tag{2}$$

where $|\psi_S(0)\rangle$ is the initial state vector, and is time-independent. This time-independent state vector is also referred to as the state vector of the Heisenberg picture, i.e., $|\psi_H\rangle = |\psi_S(0)\rangle$. The operator $\widehat{U}(t)$ is a time-evolution operator with unitary property, where $\widehat{U}^\dagger(t)\widehat{U}(t) = I$ [17]. Formally, a time evolution operator is defined as follows [17].

$$\hat{U}(t) = \exp\left(-\frac{i\hat{H}t}{\hbar}\right) \tag{3}$$

In the phase space method implemented here, the Schrödinger Eq. (1) is first expressed in terms of a density matrix ($\hat{\rho}$), the definition of which is given by [18]

$$\hat{\rho}(t) = |\psi_S(t)\rangle\langle\psi_S(t)| \tag{4}$$

The propagation of $\hat{\rho}$ can be derived by taking the first derivative of Eq. (4) as follow

$$\begin{aligned}\frac{\partial\hat{\rho}(t)}{\partial t} &= \frac{\partial}{\partial t}|\psi_S(t)\rangle\langle\psi_S(t)| \\ &= \left(\frac{\partial}{\partial t}|\psi_S(t)\rangle\right)\langle\psi_S(t)| + |\psi_S(t)\rangle\left(\frac{\partial}{\partial t}\langle\psi_S(t)|\right) \\ &= \frac{i}{\hbar}\left[|\psi_S(t)\rangle\langle\psi_S(t)|\hat{H} - \hat{H}|\psi_S(t)\rangle\langle\psi_S(t)|\right] \\ &= \frac{i}{\hbar}\left[\hat{\rho}(t)\hat{H} - \hat{H}\hat{\rho}(t)\right]\end{aligned}$$

Hence:

$$\frac{\partial\hat{\rho}(t)}{\partial t}=\frac{i}{\hbar}\left[\hat{\rho}(t),\hat{H}\right] \tag{5}$$

The above equation is known as the Liouville-Von Neumann equation, and it will be used to describe the evolution of the system in the phase space method. In fact, this equation has a solution in the form of $\hat{\rho}(t) = \hat{U}(t)\hat{\rho}(0)\hat{U}^{\dagger}(t)$ [19]. For additional information, the phase space method has been successfully employed in Refs. [20-30].

In the Heisenberg picture, the description of quantum mechanical phenomena is opposite to Schrödinger's picture. Here, the operator is said to be time-dependent, while the state vector is now fixed with time [31]. In this picture, the state vector is specified by the following expression:

$$|\psi_H\rangle=\hat{U}^{\dagger}(t)|\psi_S(t)\rangle \tag{6}$$

The above equation is obtained by algebraically modifying Eq. (2), where $|\psi_H\rangle = |\psi(0)\rangle$ is a time-independent state vector. The propagation of the operator $\hat{a}_H$ in the Heisenberg picture can be derived by taking the expectation value of $\hat{O}_S$ in the state $|\psi_S(t)\rangle$ [16].

$$\begin{aligned}\langle\psi_S(t)|\hat{O}_S|\psi_S(t)\rangle &= \langle\psi_S(0)|e^{i\hat{H}t/\hbar}\hat{O}_S e^{-i\hat{H}t/\hbar}|\psi_S(0)\rangle \\ &= \langle\psi_H|e^{i\hat{H}t/\hbar}\hat{O}_S e^{-i\hat{H}t/\hbar}|\psi_H\rangle \\ &= \langle\psi_H|\hat{O}_H|\psi_H\rangle\end{aligned}$$

where $\hat{O}_S$ is the operator of the Schrödinger picture. Therefore, the time-evolution of the operator $\hat{O}_H$ can be expressed as follows:

$$\hat{O}_H=e^{i\hat{H}t/\hbar}\hat{O}_S e^{-i\hat{H}t/\hbar} \quad \text{or} \quad \hat{O}_H(t)=\hat{U}^{+}(t)\hat{O}_S(0)\hat{U}(t) \tag{7}$$

The above equation is also called the Heisenberg operator equation, where $\hat{O}_S$ is the time-independent operator. Previously in the Schrödinger picture, the propagation of the state vector must satisfy the Schrödinger equation. Meanwhile, in the Heisenberg picture, the operator Eq. (7) is controlled by the

Heisenberg equation of motion, which is derived by taking the first derivative of Eq. (7). Hence, yielding:

$$\frac{\partial \hat{O}_H}{\partial t} = \frac{i}{\hbar}\left[\hat{H}, \hat{O}_H\right] \tag{8}$$

The above equation will be used later, to characterize the dynamics of the system using the analytical method. The details of the applications will be further explained in the next section.

2. Mathematical Formulation

For the considered system, we recall the Hamiltonian operator used in the previous chapter (Eq. 3) for two-channel system as follows:

$$\begin{aligned}\hat{H} = \hbar\Big[&\omega_1\hat{a}_1^{\dagger}\hat{a}_1 + \omega_2\hat{a}_2^{\dagger}\hat{a}_2 + (2\omega_1)\hat{b}_1^{\dagger}\hat{b}_1 + (2\omega_2)\hat{b}_2^{\dagger}\hat{b}_2 \\ &+ \frac{ig}{2}\left(\hat{a}_1^{\dagger 2}\hat{b}_1 + \hat{a}_2^{\dagger 2}\hat{b}_2 - \text{h.c.}\right) + \kappa\left(\hat{a}_1^{\dagger}\hat{a}_2 + \text{h.c.}\right)\Big]\end{aligned} \tag{9}$$

In the above Hamiltonian, the term with ω_j and $2\omega_j$
propagation of the fundamental and second harmonic modes, respectively, where the parameter ω_j
the term with g and κ specify the nonlinear coupling between the fundamental mode and second harmonic mode, and the linear coupling between the fundamental modes, accordingly.

2.1. Phase Space Method for Contra-Directional Two-Channel System

In the phase space method, the system evolution is described by inserting the Hamiltonian Eq. (9) into the Von-Neumann Eq. (5). Using the standard procedure in quantum mechanics, the quantum mechanical master equation of the density matrix can be derived as explained in the previous chapter. This density matrix quantum mechanical equation is then converted to its corresponding classical Fokker-Planck equation via the operator correspondences of the positive-P representation (See Eqs. 6-9 in chapter 6). Via Ito calculus rules, the Fokker Planck equation is mapped into the set of Langevin Stochastics differential equations with Gaussian noise terms where the magnitude of the noise is given by the square-root of the diffusion matrix. This procedure yields the same stochastic system of differential equations obtained in the previous chapter (See Eqs. 11-18 in chapter 6):

$$\frac{d\alpha_1}{dt} = -i\omega_1\alpha_1 + g\bar{\alpha}_1\beta_1 - i\kappa\alpha_2 + \sqrt{g\bar{\alpha}_1}\eta_1(t) \tag{10}$$

$$\frac{d\beta_1}{dt} = i\omega_1\beta_1 + g\alpha_1\bar{\beta}_1 + i\kappa\beta_2 + \sqrt{g\bar{\beta}_1}\eta_2(t) \tag{11}$$

$$\frac{d\alpha_2}{dt} = -i\omega_2\alpha_2 + g\bar{\alpha}_2\beta_2 - i\kappa\alpha_1 + \sqrt{g\bar{\alpha}_2}\eta_3(t) \tag{12}$$

$$\frac{d\beta_2}{d\tau} = i\omega_2\beta_2 + g\alpha_2\bar{\beta}_2 + i\kappa\beta_1 + \sqrt{g\bar{\beta}_2}\eta_4(t) \tag{13}$$

$$\frac{d\bar{\alpha}_1}{dt} = -i(2\omega_1)\bar{\alpha}_1 - \frac{g}{2}\alpha_1^2 \tag{14}$$

$$\frac{d\bar{\beta}_1}{dt} = i(2\omega_1)\bar{\beta}_1 - \frac{g}{2}\beta_1^2 \tag{15}$$

$$\frac{d\bar{\alpha}_2}{dt} = -i(2\omega_2)\bar{\alpha}_2 - \frac{g}{2}\alpha_2^2 \tag{16}$$

$$\frac{d\bar{\beta}_2}{dt} = i(2\omega_2)\bar{\beta}_2 - \frac{g}{2}\beta_2^2 \tag{17}$$

With the existence of noise terms in the above system, a single trajectory solution becomes meaningless. The outcome of a single solution represents the fluctuation of noise in the system. To obtain a stable solution, the solution has to be averaged over many trajectories. The spatial form of the stochastic differential equations can be expressed as:

$$\frac{d\alpha_1}{dz} = -ik_1\alpha_1 + G\bar{\alpha}_1\beta_1 - iK\alpha_2 + \sqrt{G\bar{\alpha}_1}\xi_1(z) \tag{18}$$

$$\frac{d\beta_1}{dz} = ik_1\beta_1 + G\alpha_1\bar{\beta}_1 + iK\beta_2 + \sqrt{G\bar{\beta}_1}\xi_2(z) \tag{19}$$

$$\frac{d\alpha_2}{dz} = -\left[-ik_2\alpha_2 + G\bar{\alpha}_2\beta_2 - iK\alpha_1 + \sqrt{G\bar{\alpha}_2}\xi_3(z)\right] \tag{20}$$

$$\frac{d\beta_2}{dz} = -\left[ik_2\beta_2 + G\alpha_2\bar{\beta}_2 + iK\beta_1 + \sqrt{G\bar{\beta}_2}\xi_4(z)\right] \tag{21}$$

$$\frac{d\bar{\alpha}_1}{dz} = -i(2k_1)\bar{\alpha}_1 - \frac{G}{2}\alpha_1^2 \tag{22}$$

$$\frac{d\bar{\beta}_1}{dz} = i(2k_1)\bar{\beta}_1 - \frac{G}{2}\beta_1^2 \tag{23}$$

$$\frac{d\bar{\alpha}_2}{dz} = -\left[-i(2k_2)\bar{\alpha}_2 - \frac{G}{2}\alpha_2^2\right] \tag{24}$$

$$\frac{d\bar{\beta}_2}{dz} = -\left[i(2k_2)\bar{\beta}_2 - \frac{G}{2}\beta_2^2\right] \tag{25}$$

where $k_j = \frac{\omega_j}{v_c}$, $K = \frac{\kappa}{v_c}$, $G = \frac{g}{v_c}$, $\xi(z) = \frac{\eta(t)}{\sqrt{v_c}}$ and v_c
coupler.

The above form of equations is necessary to study the contra-directional propagation system since the time (t) is non-directional in nature. A contra-directional mode can be introduced by inserting a negative (–) sign in the equation of motion for one of the two modes. That is, the mode propagates in the negative direction relative to a cartesian coordinate system. The negative sign is inserted into Eqs. (20)–(21) and (24)–(25) so that the mode $\hat{a}_2$ can travel in the opposite direction.

In addition, Eqs. (18)–(25) are then scaled into a unitless form to ensure convenience in computational simulations. In a unitless form, we use the dimensionless parameters that can be expressed in terms of their dimensional counterparts as $\tilde{k}_1 = \frac{k_1}{k_1} = 1$, $\tilde{k}_2 = \frac{k_2}{k_1}$, $\tilde{G} = \frac{G}{k_1}$, $\tilde{K} = \frac{K}{k_1}$, $\tilde{z} = k_1 z$ and $\tilde{\xi}(\tilde{z}) = \frac{\tilde{\xi}(z)}{\sqrt{k_1}}$

2.2. Analytical Method for Contra-Directional Two-Channel System

Using the same technique as in Chapter 6, the Heisenberg equations of motion for the current system can be obtained by evaluating the general form of the Heisenberg equation of motion (Eq. (8)) using the Hamiltonian given by Eq. (9). This gives

$$\frac{d\hat{a}_1}{dz} = -ik_1\hat{a}_1 + G\hat{a}_1{}^\dagger\hat{b}_1 - iK\hat{a}_2 \tag{26}$$

$$\frac{d\hat{a}_2}{dz} = -\left[-ik_2\hat{a}_2 + G\hat{a}_2{}^\dagger\hat{b}_2 - iK\hat{a}_1\right] \tag{27}$$

$$\frac{d\hat{b}_1}{dz} = -i(2k_1)\hat{b}_1 - \frac{G}{2}\hat{a}_1{}^2 \tag{28}$$

$$\frac{d\hat{b}_2}{dz} = -\left[-i(2k_2)\hat{b}_2 - \frac{G}{2}\hat{a}_2{}^2\right] \tag{29}$$

The former set of equations (26)-(29) are similar to Eqs. (20)-(23) in the previous chapter. However, Eqs. (26)-(29) are now expressed in term of spatial parameter z rather than time by defining the new set of coefficients $k_j = \frac{\omega_j}{v_c}$, $K = \frac{\kappa}{v_c}$ and $G = \frac{g}{v_c}$. Furthermore, a negative sign at the outer bracket of Eqs. (27) and (29) are added to introduce the contra-directional mode propagation in the second channel. By using the Heisenberg equation of motion in the form of Eqs. (26)–(29), the commutator terms in $\hat{a} = \hat{a}(0) + \frac{iv_c t}{\hbar}\left[\frac{\hat{H}}{v_c}, \hat{a}(0)\right] - \frac{1}{2}\frac{v_c^2 t^2}{\hbar^2}\left[\frac{\hat{H}}{v_c}, \left[\frac{\hat{H}}{v_c}, \hat{a}(0)\right]\right]$ are evaluated, yielding the assumed solution with the combination of several terms as follows:

$$\hat{a}_1(L)=U_1\hat{a}_1(0)+U_2\hat{a}_2(0)+U_3\hat{a}_1^{\dagger}(0)\hat{b}_1(0)$$
$$+U_4\hat{a}_2^{\dagger}(0)\hat{b}_1(0)+U_5\hat{a}_2^{\dagger}(0)\hat{b}_2(0)$$
$$+U_6\hat{a}_1(0)\hat{b}_1^{\dagger}(0)\hat{b}_1(0)+U_7\hat{a}_1^{\dagger}(0)\hat{a}_1^{2}(0) \tag{30}$$

$$\hat{a}_2(0)=V_1\hat{a}_1(L)+V_2\hat{a}_2(L)+V_3\hat{a}_2^{\dagger}(L)\hat{b}_2(L)$$
$$+V_4\hat{a}_1^{\dagger}(L)\hat{b}_2(L)+V_5\hat{a}_1^{\dagger}(L)\hat{b}_1(L)$$
$$+V_6\hat{a}_2^{\dagger}(L)\hat{a}_2^{2}(L)+V_7\hat{a}_2(L)\hat{b}_2^{\dagger}(L)\hat{b}_2(L) \tag{31}$$

$$\hat{b}_1(L)=\bar{U}_1\hat{b}_1(0)+\bar{U}_2\hat{a}_1^{2}(0)+\bar{U}_3\hat{a}_2(0)\hat{a}_1(0)$$
$$+\bar{U}_4\hat{a}_1^{\dagger}(0)\hat{a}_1(0)\hat{b}_1(0)+\bar{U}_5\hat{a}_1(0)\hat{a}_1^{\dagger}(0)\hat{b}_1(0) \tag{32}$$

$$\hat{b}_2(0)=\bar{V}_1\hat{b}_2(L)+\bar{V}_2\hat{a}_2^{2}(L)+\bar{V}_3\hat{a}_1(L)\hat{a}_2(L)$$
$$+\bar{V}_4\hat{a}_2^{\dagger}(L)\hat{a}_2(L)\hat{b}_2(L)+\bar{V}_5\hat{a}_2(L)\hat{a}_2^{\dagger}(L)\hat{b}_2(L) \tag{33}$$

where we have put $z = L$ in the previous system (30)-(33). For further details on the of contradirectional propagation process, the reader is referred to J. Perina work (e.g., [32]) on contradirectional propagation based on the short-term approximation method. Substituting the proposed solutions (Eqs. (30)–(33)) into the Heisenberg equations of motion (26) and equating the similar terms on both sides, a set of differential equations can be obtained. These differential equations specify the spatial evolution of the functions $\{U_k\}_1^7$, $\{V_k\}_1^7$, $\{\bar{U}_k\}_1^5$ and $\{\bar{V}_k\}_1^5$. For U_k, the equations are given as follows:

$$\frac{dU_1}{dz}=-i\tilde{k}_1U_1-i\tilde{K}V_1 \tag{34}$$

$$\frac{dU_2}{dz}=-i\tilde{k}_1U_2-i\tilde{K}V_2 \tag{35}$$

$$\frac{dU_3}{dz}=-i\tilde{k}_1U_3+\tilde{G}U_1^{*}\bar{U}_1-i\tilde{K}V_5 \tag{36}$$

$$\frac{dU_4}{dz} = -i\tilde{k}_1 U_4 + \tilde{G} U_2^* \bar{U}_1 \tag{37}$$

$$\frac{dU_5}{dz} = -i\tilde{k}_1 U_5 - i\tilde{K} V_3 \tag{38}$$

$$\frac{dU_6}{dz} = -i\tilde{k}_1 U_6 + \tilde{G} U_3^* \bar{U}_1 \tag{39}$$

$$\frac{dU_7}{dz} = -i\tilde{k}_1 U_7 + G U_1^* \bar{U}_2 \tag{40}$$

Repeating similar steps

Now Substituting the proposed solutions in Eqs. (30) – (33) into the Heisenberg equations of motion (27) and equating the similar terms on both sides, a set of differential equations for V_k

$$\frac{dV_1}{dz} = -\left[-i\tilde{k}_2 V_1 - i\tilde{K} U_1\right] \tag{41}$$

$$\frac{dV_2}{dz} = -\left[-i\tilde{k}_2 V_2 - i\tilde{K} U_2\right] \tag{42}$$

$$\frac{dV_3}{dz} = -\left[-i\tilde{k}_2 V_3 + \tilde{G} V_2^* \bar{V}_1 - i\tilde{K} U_5\right] \tag{43}$$

$$\frac{dV_4}{dz} = -\left[-i\tilde{k}_2 V_4 + \tilde{G} V_1^* \bar{V}_1\right] \tag{44}$$

$$\frac{dV_5}{dz} = -\left[-i\tilde{k}_2 V_5 - i\tilde{K} U_3\right] \tag{45}$$

$$\frac{dV_6}{dz} = -\left[-i\tilde{k}_2 V_6 + \tilde{G} V_2^* \bar{V}_2\right] \tag{46}$$

$$\frac{dV_7}{dz} = -\left[-i\tilde{k}_2 V_7 + \tilde{G} V_3^* \bar{V}_1\right] \quad (47)$$

In similar fashion, substituting the proposed solutions in Eqs. (30) – (33) into the Heisenberg equations of motion (27)-(28) and equating the similar terms on both sides, a set of differential equations for $\bar{U}_k$ and $\bar{V}_k$ are obtained as below;

$$\frac{d\bar{U}_1}{dz} = -i\left(2\tilde{k}_1\right)\bar{U}_1 \quad (48)$$

$$\frac{d\bar{U}_2}{dz} = -i\left(2\tilde{k}_1\right)\bar{U}_2 - \frac{\tilde{G}}{2}U_1^2 \quad (49)$$

$$\frac{d\bar{U}_3}{dz} = -i\left(2\tilde{k}_1\right)\bar{U}_3 - \tilde{G}U_1U_2 \quad (50)$$

$$\frac{d\bar{U}_4}{dz} = -i\left(2\tilde{k}_1\right)\bar{U}_4 - \tilde{G}U_3U_1 \quad (51)$$

$$\frac{d\bar{U}_5}{dz} = -i\left(2\tilde{k}_1\right)\bar{U}_5 - \tilde{G}U_1U_3 \quad (52)$$

$$\frac{d\bar{V}_1}{dz} = -\left[-i\left(2\tilde{k}_2\right)\bar{V}_1\right] \quad (53)$$

$$\frac{d\bar{V}_2}{dz} = -\left[-i\left(2\tilde{k}_2\right)\bar{V}_2 - \frac{\tilde{G}}{2}V_2^2\right] \quad (54)$$

$$\frac{d\bar{V}_3}{dz} = -\left[-i\left(2\tilde{k}_2\right)\bar{V}_3 - \tilde{G}V_1V_2\right] \quad (55)$$

$$\frac{d\bar{V}_4}{dz} = -\left[-i\left(2\tilde{k}_2\right)\bar{V}_4 - \frac{\tilde{G}}{2}V_3V_2\right] \quad (56)$$

$$\frac{d\bar{V}_5}{dz} = -\left[-i\left(2\tilde{k}_2\right)\bar{V}_5 - \frac{\tilde{G}}{2}V_2V_3 \right] \tag{57}$$

respectively. These equations are solved by using the appropriate mathematical software, and the solutions are used to form the absolute mode operator solutions. Since the calculation of quadrature variances requires the operation between the expectation value of creation and annihilation operator of modes, the obtained mode solution can directly be used in the investigation of single-mode squeezing. For additional information, the analytical method that we used here is superior to the short-length approximation method, that is only suitable for short investigation period only. Using the short-length approximation, squeezing is not too accurate, since no oscillating energy is detected (see ref. [33]). Hence, the method employed here can be an alternative to the short-length approximation method.

2.3. Squeezing in Contra-Directional Two-Channel System

The single-mode squeezing is evaluated using the quadrature variances formulas acquired in the preceding chapter for the phase space technique (See Eq. 69). These are

$$\left\langle \left(\Delta\hat{X}\right)^2 \right\rangle = \frac{1}{4}\left(1 + 2\langle\beta\alpha\rangle - 2\langle\alpha\rangle\langle\beta\rangle + \langle\alpha^2\rangle + \langle\beta^2\rangle - \langle\alpha\rangle^2 - \langle\beta\rangle^2\right) \tag{58}$$

$$\left\langle \left(\Delta\hat{Y}\right)^2 \right\rangle = \frac{1}{4}\left(1 + 2\langle\beta\alpha\rangle - 2\langle\alpha\rangle\langle\beta\rangle - \langle\alpha^2\rangle - \langle\beta^2\rangle + \langle\alpha\rangle^2 + \langle\beta\rangle^2\right) \tag{59}$$

For the analytical approach, the mode operator solutions (Eqs. (30) – (33)) are substituted into the following expressions for the quadrature variances;

$$\left\langle \left(\Delta\hat{X}\right)^2 \right\rangle = \frac{1}{4}\left(1 + 2\langle\hat{a}^\dagger\hat{a}\rangle - 2\langle\hat{a}\rangle\langle\hat{a}^\dagger\rangle + \langle\hat{a}^2\rangle + \langle\hat{a}^{\dagger 2}\rangle - \langle\hat{a}\rangle^2 - \langle\hat{a}^\dagger\rangle^2\right) \tag{60}$$

$$\left\langle \left(\Delta\hat{Y}\right)^2 \right\rangle = \frac{1}{4}\left(1 + 2\langle\hat{a}^\dagger\hat{a}\rangle - 2\langle\hat{a}\rangle\langle\hat{a}^\dagger\rangle - \langle\hat{a}^2\rangle - \langle\hat{a}^{\dagger 2}\rangle + \langle\hat{a}\rangle^2 + \langle\hat{a}^\dagger\rangle^2\right) \tag{61}$$

yielding the following expressions:

$$\left\langle \left(\Delta\hat{X}_{a_1}\right)^2\right\rangle, \left\langle \left(\Delta\hat{Y}_{a_1}\right)^2\right\rangle = \frac{1}{4}\Big[1+2\Big(|U_5|^2|\beta_2|^2+U_5U_4^*\beta_1^*\beta_2$$
$$+U_4U_5^*\beta_2^*\beta_1\Big)\pm\Big(U_3U_1\beta_1+U_7U_1\alpha_1^2$$
$$+U_4U_2\beta_1+U_5U_2\beta_2+\text{c.c.}\Big)\Big] \quad (62)$$

$$\left\langle \left(\Delta\hat{X}_{a_2}\right)^2\right\rangle, \left\langle \left(\Delta\hat{Y}_{a_2}\right)^2\right\rangle = \frac{1}{4}\Big[1+2\Big(|V_5|^2|\beta_1|^2+V_5V_4^*\beta_2^*\beta_1$$
$$+V_4V_5^*\beta_1^*\beta_2\Big)\pm\Big(V_4V_1\beta_2+V_5V_1\beta_1$$
$$+V_3V_2\beta_2+V_6V_2\alpha_2^2+\text{c.c.}\Big)\Big] \quad (63)$$

Note that all these expressions are obtained by considering the normal-ordered operator arrangement. Otherwise, a meaningless outcome will be generated. The term c.c. in the above equations symbolize the complex conjugate term, while α_1, α_2, β_1 and β_2 amplitude of the mode $\hat{a}_1(0)$, $\hat{a}_2(0)$, $\hat{b}_1(0)$ and $\hat{b}_2(0)$, respectively. All these equations are used to plot the evolution of quadrature variances to identify the occurrence of squeezing in the particular system.

3. Results and Discussions

To understand more on the nature of squeezing in the considered system, the function of quadrature variances obtained in the previous section is plotted against the modulus of interaction distance $|z|$. In Figure 2, the squeezing evaluated for the first mode, $\hat{a}_1$ (Figure 2(a)) and the second mode, $\hat{a}_2$ (Figure 2(b)) are presented. For the first mode, both methods predict squeezing with a steady pattern, and both of them show the increment of squeezing amplitude over distance. For the second mode, a similar pattern is generated, maybe due to the symmetrical amplitude initialization. However, for the analytical method, if the graph lies below the shot noise for codirectional mode, the graph will lie above the shot noise for the contra-directional mode, and vice versa, and this may be due to its direction that is in opposite to $\hat{a}_1$. Nonetheless, this observation is not obtained in the phase space result, where both the codirectional and contra-directional modes generate a similar graph. We can also see in Figure 2(b) that the first quadrature component of the analytical method agrees with the second quadrature component of the phase space method, and vice versa. Therefore, no agreement is achieved in the contra-directional mode.

Since the phase space method and the analytical method do not agree for the mode $\hat{a}_2$, the other effect will be discussed by considering the mode $\hat{a}_1$ only. Now, we will focus on the effect of linear coupling interaction. We use $\widetilde{K} = 0.04$ in Figure 3(a) and $\widetilde{K} = 0.4$ in Figure 3(b), where these two values are set similarly to ref. [34]. When $\widetilde{K} = 0.04$, the pattern is similar to the previous result in Figure 2(a), in which both the methods oscillate steadily, but now, the agreement between both the predictions is higher. From this effect, we can say that both the methods will less agree to each other when strong coupling is applied. This is clearer if we observe Figure 3(b), where the disagreement between them is more obvious.

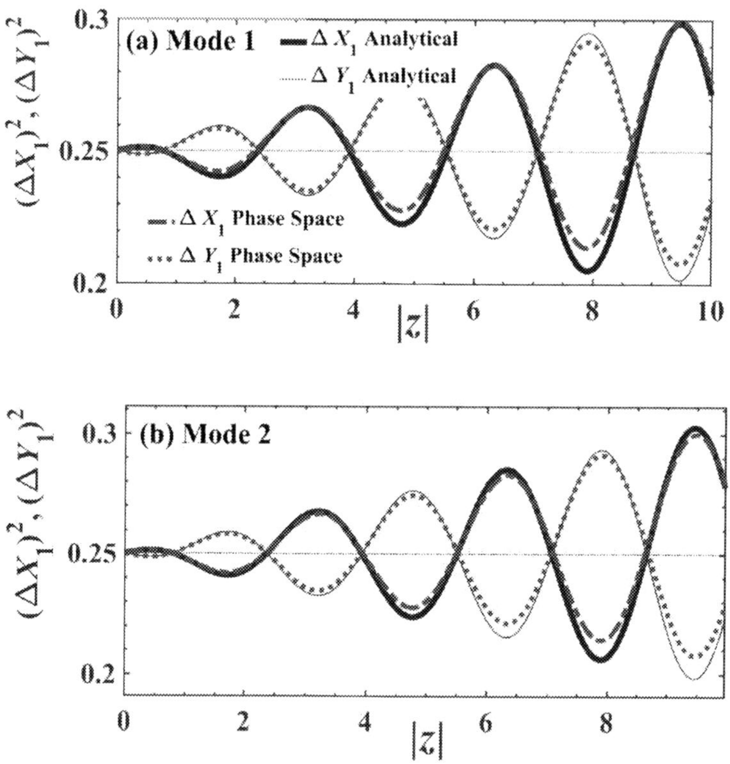

Figure 2. Variation of squeezing in (a) mode $\hat{a}_1$, and (b) mode $\hat{a}_2$ for input parameters $\alpha_1 = \alpha_2 = 1$, $\tilde{k}_1 = \tilde{k}_2 = 1$, $\tilde{G} = 0.01$ and $\widetilde{K} = 0.1$.

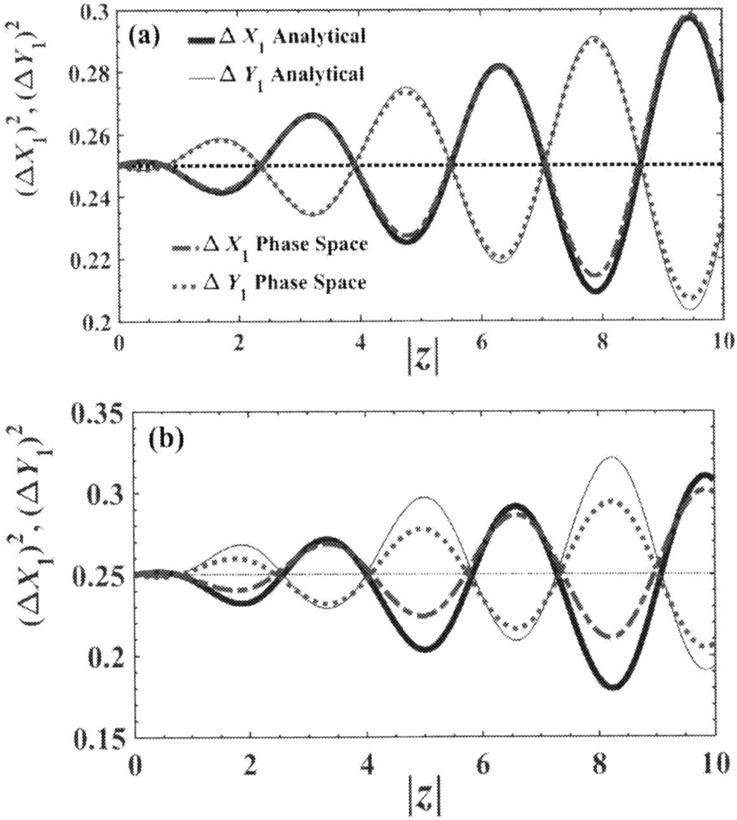

Figure 3. Fluctuation of squeezing for the first mode $\hat{a}_1$ with (a) $\widetilde{K} = 0.04$, and (b) $\widetilde{K} = 0.4$. Other parameters are $\alpha_1 = \alpha_2 = 1$, $\tilde{k}_1 = \tilde{k}_2 = 1$ and $\tilde{G} = 0.01$.

In Figure 4, the squeezing generated using different nonlinear coupling coefficients is presented.

Here, we compare the situation, where $\tilde{G} = 0.011$ (Figure 4(a)) with the situation, where $\tilde{G} = 0.1$ (see Figure 4(b)). When $\tilde{G} = 0.011$, both the predictions seem fine, but when $\tilde{G} = 0.1$, the phase space method becomes unstable starting at $|z|{\sim}8$. Hence, the result for the range beyond $|z|{\sim}8$ is not known. But, we predict that the squeezing signal losess its energy and disappeared after that moment. From this result, we can conclude that the phase space method is unstable for high nonlinear coupling and long interaction distance, but its capability to detect the disappearance of squeezing signal makes it a more realistic method.

The variation of squeezing under the phase mismatch condition is shown in Figure 5. Phase mismatch is the condition where the mode with a different wavenumber is introduced. We plot the squeezing with the phase mismatch of $\Delta\tilde{k} = 1.5$ and $\Delta\tilde{k} = 4$ in Figure 5(a)–(b), respectively. From our observation in Figure 5(a), collapses-revivals-like pattern is obtained when the phase mismatch is introduced in the system. This pattern is significant in the analytical method, but less significant in the phase space prediction. When the phase mismatch is increased (see Figure 5(b)), both the methods produce a steady pattern just like when there is no phase mismatch in the system. Also, the agreement between these two predictions is high at this moment, especially in the region above the shot noise.

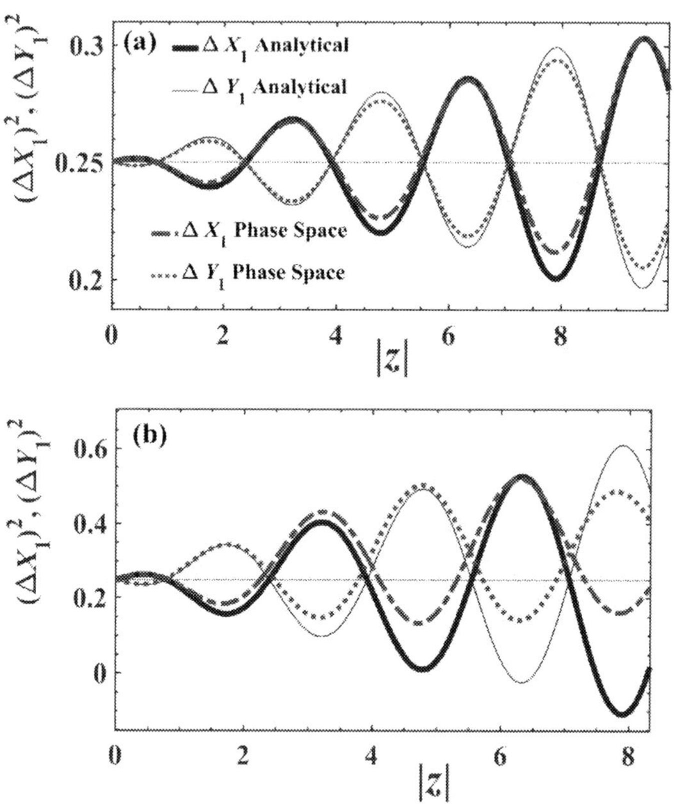

Figure 4. Evolution of quadrature variances in the first mode $\hat{a}_1$ with (a) $\tilde{G} = 0.011$, and (b) $\tilde{G} = 0.1$. Other parameters are $\alpha_1 = \alpha_2 = 1$, $\tilde{k}_1 = \tilde{k}_2 = 1$ and $\tilde{K} = 0.1$.

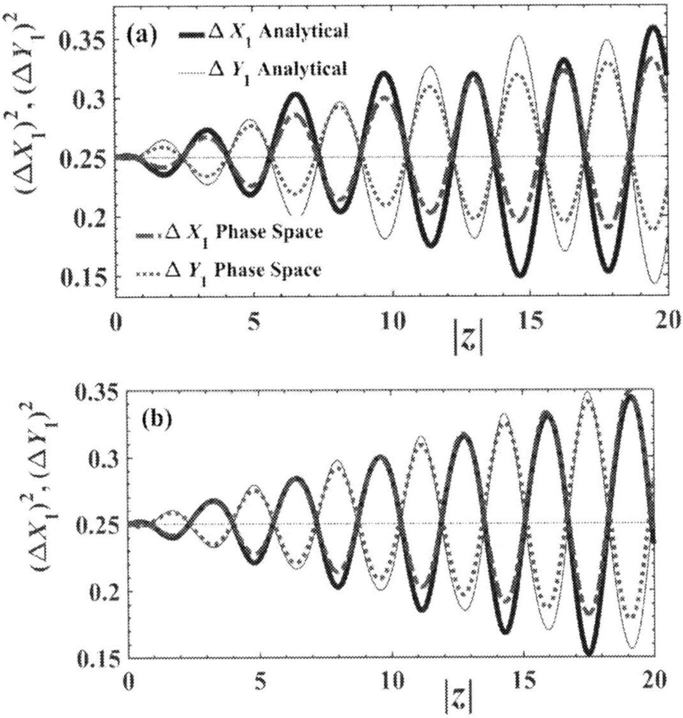

Figure 5. Squeezing evolution in the first mode $\hat{a}_1$ with phase mismatch, (a) $\Delta\tilde{k} = 1.5$, and (b) $\Delta\tilde{k} = 4$. Other parameters are $\alpha_1 = \alpha_2 = 1, \tilde{G} = 0.01$ and $\widetilde{K} = 0.4$.

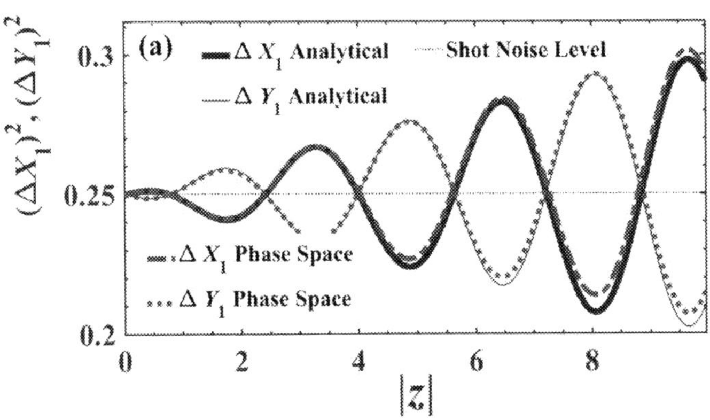

Figure 6. Squeezing evolution in (a) the first mode $\hat{a}_1$, and (b) the second mode $\hat{a}_2$ under the asymmetrical amplitude initialization, where $\alpha_1 = 1$ and $\alpha_2 = 0.01$. (a) $\Delta\tilde{k} = 1.5$, and (b) $\Delta\tilde{k} = 4$. Other parameters are $\tilde{k}_1 = \tilde{k}_2 = 1, \tilde{G} = 0.01$ and $\widetilde{K} = 0.3$.

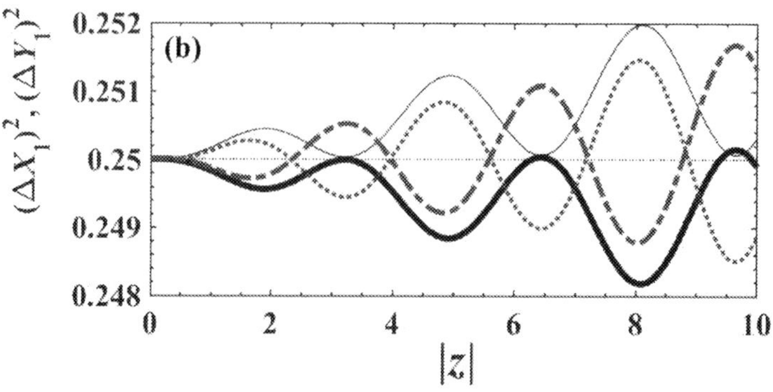

Figure 6. (Continued)

In Figure 6, the graphs are plotted under the asymmetrical initialization, where the amplitude of the second mode is set to be nearly equal to vacuum. We cannot set $\alpha_2 = 0$ (pure vacuum state) for the contra-directional system, since the vacuum state has no direction. Hence, we use $\alpha_2 = 0.01$ to make it valid with the contradirectional propagation system. The results for the first and the second mode are plotted in Figure 6(a)–(b), respectively. Previously, in Figure 2(b), the phase space method disagreed with the analytical method for the second (contra-directional) mode, when symmetrical initialization is applied. But here, under asymmetrical initialization, a slight agreement between both methods is achieved, since both predictions evolve through a similar path. However, this agreement is maintained for short interaction distance (up to $\tau \sim 1$) only.

Conclusion

Finally, squeezing in a two-channel contra-directional nonlinear coupler with SHG has been studied using the phase space approach based on the Schrodinger picture and the analytical method based on the Heisenberg picture of quantum mechanics. These two methods are mostly used to get quadrature squeezing variances from solutions for mode operators. When no phase mismatch is added, both approaches anticipate a steady oscillation pattern. The phase space technique does not agree with the analytical method for the second (contra-directional) mode under symmetrical initialization, although there is a modest agreement with asymmetrical initialization. The peak squeezing amplitude increases as the linear coupling coefficient increases, but the squeezing oscillation gets less frequent. Furthermore, when the linear coupling improves, the agreement between the two techniques decreases. When the nonlinear coupling is high, the positive-P phase space approach fails to identify squeezing at a long interaction distance. However, because squeezing attenuation is frequently noticed, the prediction by the phase space method seems more reasonable. Finally, when there is a phase mismatch, a collapses-revivals pattern is created. This collapses-revivals pattern returns to its original steady oscillation pattern when the phase mismatch increases.

Acknowledgments

Abdel-Baset M. A. Ibrahim gratefully recognizes the Malaysian Ministry of Higher Education (MOHE) national grant FRGS/1/2021/STG07/UITM/02/10 and the internal grant 600-RMC/MYRA 5/3/LESTARI (069/2020) from Universiti Teknologi MARA (UiTM).

References

[1] Schnabel R. "Squeezed states of light and their applications in laser interferometers", *Phys Rep.* vol. 684, pp 1-51, 2017.

[2] Mehmet M and Vahlbruch H. "The squeezed light source for the Advanced Virgo detector in the observation run O_3", *Galaxies.* vol 8, no 4, p 79, 2020.

[3] Alaskary SD and Fyath RS. "Binary quantum communication using squeezed light: Numerical, simulation, and experimental results", p 21, *Int J Comput Sci Inf Technol.* vol 11 no 7, pp 2839-2858, 2013.

[4] Xu C et al. "Sensing and tracking enhanced by quantum squeezing", *Photon Res.* vol 7, no 6, p A14, 2019.

[5] Wu B-H, Alexander RN, Liu S and Zhang Z. "Quantum computing with multidimensional continuous-variable cluster states in a scalable photonic platform", *Phys Rev Research.* vol 2, no 2, p 023138, 2020.

[6] Yonezawa H, Braunstein SL and Furusawa A. "Experimental demonstration of quantum teleportation of broadband squeezing", *Phys Rev Lett.* vol 99, no 11, p 110503, 2007.

[7] Lie SH and Jeong H. "Limitations of teleporting a qubit via a two-mode squeezed state", *Photon Res.* vol 7, no 5, p A7, 2019.

[8] Adhikari S, Majumdar AS and Nayak N. "Teleportation of two-mode squeezed states", *Phys Rev A.* vol 77, no 1, p 012337, 2008.

[9] Kumar Choudhary B and Giri DK. "Squeezing, sub-Poissonian and total noise in degenerate six-wave mixing process," *Acta Phys Pol A*. vol 134, no 6, pp 1108–1114, 2018.

[10] Kryuchkyan GY and Kheruntsyan KV. "Squeezing in intracavity four-wave mixing with non-degenerate pumps", *Quantum Opt.* vol 4, no 5, pp 289-302, 1992.

[11] Feng L-J, Lin G-W, Deng L, Niu Y-P and Gong S-Q. "Strong mechanical squeezing in an electromechanical system", *Sci Rep*. vol 8, no 1, p 3513, 2018.

[12] Sayed Hassen SZ. "Modelling, analysis and simulation of an optical squeezer". *Appl Math Modell.* vol 39, no 13, pp 3846–3861, 2015.

[13] Albano L, Mundarain DF and Stephany J. "On the squeezed number states and their phase space representations", *J Opt B: Quantum Semiclass Opt.* vol 4, no 5, pp 352–357, 2002.

[14] Iida S, Yukawa M, Yonezawa H, Yamamoto N and Furusawa A. "Experimental demonstration of coherent feedback control on optical field squeezing", *IEEE Trans Automat Contr.* vol 57, no 8, pp 2045–2050, 2012.

[15] Vahlbruch H, Mehmet M, Danzmann K and Schnabel R. "Detection of 15 dB squeezed states of light and their application for the absolute calibration of photoelectric quantum efficiency", *Phys Rev Lett.* vol 117, no 11, p 110801, 2016.

[16] Zettili N. "*Quantum Mechanics: Concepts and Applications*", 2nd ed. Wiley, Chichester, UK (2009).

[17] Enríquez M and Cruz y Cruz S. "Disentangling the Time-Evolution Operator of a Single Qubit", *J Phys: Conf Ser.* vol 839, p 012015, 2017.

[18] Ficek Z and Wahiddin MR. "*Quantum Optics for Beginners*", 1st ed. Pan Stanford Publishing, Singapore (2016).

[19] Czachor M, Doebner H-D, Syty M and Wasylka K. "von Neumann equations with time-dependent Hamiltonians and supersymmetric quantum mechanics", *Phys Rev E.* vol 61, no 4, pp 3325–3329, 2000.

[20] Ibrahim A-B MA, Umarov BA and Wahiddin MRB. "Squeezing in the Kerr nonlinear coupler via phase-space representation", *Phys Rev A*. vol 61, no 4, p 043804, 2000.

[21] Umarov BA, Ibrahim A-B MA, Wahiddin MRB. Quantum Dynamics in Kerr Nonlinear Coupler Via Phase-Space Representation. In "*Nonlinearity and Disorder: Theory and Applications*". Ed. F Abdullaev, O Bang, M. Sørensen, NATO Science Series (Series II: Mathematics, Physics and Chemistry), vol 45, pp 341-349 Springer, Dordrecht (2001).

[22] Julius R, Ibrahim A-B MA and Deni MSM. "Quantum state generation in a four-mode Kerr nonlinear directional coupler". *Laser Phys.* vol 24, no 3, p 035202, 2014.

[23] Julius R and Ibrahim A-B MA. "Quantum squeezing in multichannel nonlinear coupler", *JCSCM*. pp 131-136, 2017.

[24] Julius R, Ibrahim A-B MA, Abd-Rahman MK et al. Quantum dynamics of a four-channel Kerr nonlinear directional coupler. *Opt Rev.* vol 25, no 5, pp. 563-570, 2018.

[25] Julius R, Ibrahim A-B MA, Choudhury PK and Eleuch H. "On the nonclassical dynamics of cavity-assisted four-channel nonlinear coupler", *Chinese Phys B*. vol 27, no 11, p 114206, 2018.

[26] Julius R, Ibrahim A-B MA and Choudhury Pankaj K. Dielectric Waveguides as a Source for Nonclassical Light. In "*Dielectric Materials and Applications*", Ed., PK Choudhury, Nova science publishers, USA, pp 335-358 (2019).

[27] Julius R, Ibrahim A-B MA, Choudhury PK and Eleuch H. "Quantum states generation in multichannel directional coupler with second-order nonlinearity", *Optik*. vol 186, pp 212-220, 2019.

[28] Julius R, Ibrahim A-B MA, Eleuch H and Choudhury PK. "Sub-Poissonian photon squeezing and entanglement in optical chain second harmonic generation", *J Mod Opt*. vol 66, no 10, pp 1129-1138, 2019.

[29] Julius R, Zahirzai M, Ibrahim A-B MA, Eleuch H and Choudhury PK. "Single- and compound-mode squeezing in nonlinear coupler with frequency mismatch", *J Electromagn Waves Appl*. vol 34, no 3, pp 301-315, 2020.

[30] Hanapi MSM, Ibrahim A-B MA, Julius R and Eleuch H. "The quantum Kerr nonlinear coupler: Analytical versus phase-space method", *Can J Phys*. 99, no 999, 832-840, 2021.

[31] Choudhary K, Musa S, Nengem SM and Choudhary SK. "Time-evolution as an operator in Hilbert spaces with applications in quantum mechanics", *IJMTT*. vol 10, no 2, pp 60-69, 2014.

[32] Perina J and Perina Jr J. "Quantum statistical properties of codirectional and contradirectional nonlinear couplers with phase mismatch", *J Eur Opt Soc Part B*. vol 7, no 5, pp 863-876, 1995.

[33] Peřina J and Bajer J. "Non-classical light in nonlinear symmetric and asymmetric couplers", *J Mod Opt*. vol 42, no 11, pp 2337-2346, 1995.

[34] Korolkova N and Peřina J. "Quantum statistics and dynamics of Kerr nonlinear couplers", *Opt Commun*. vol 136, no 1-2, pp 135-149, 1997.

Chapter 8

Squeezed Light in Coupled Waveguide Networks Induced by the Nonlinear Kerr Effect and Second Harmonic Generation

Rafael Julius[1], Abdel-Baset M. A. Ibrahim[2,*] and Pankaj K. Choudhury[3]
[1]Faculty of Applied Sciences, Universiti Teknologi MARA (UiTM) Perak, Tapah Campus, Perak, Malaysia
[2]Faculty of Applied Sciences, Universiti Teknologi MARA (UiTM), Shah Alam, Selangor, Malaysia
[3]Institute of Microengineering and Nanoelectronics, Universiti Kebangsaan Malaysia, UKM Bangi, Selangor, Malaysia

Abstract

This chapter shows that coupled waveguide arrays with combined second and third-order nonlinear effects can generate and enhance nonclassical light in single and multi-mode settings. On the generation of squeezed light, we investigate the effects of cross-action evanescent coupling between the channel waveguides with different order of nonlinearities, the number of second harmonic generation waveguides, mismatched frequencies, and compound-mode interaction. We look at how squeezing behaves and how coupled nonlinearity affects its evolution. We demonstrate that combined nonlinear effects in coherently driven coupled waveguide networks can be a valuable source of squeezed light.

* Corresponding Author's Email: abdelbaset@uitm.edu.my.

In: Second Harmonic Generation
Editors: Abdel-Baset M. A. Ibrahim and Pankaj Kumar Choudhury
ISBN: 978-1-68507-888-1

Keywords: Quantum optics, second harmonic generation, Kerr nonlinearity, Kerr effect, optical coupler, squeezed states, squeezing, second harmonic generation, phase space representation, nonlinear waveguides

1. Introduction

Integrated quantum photonics engineering is at the heart of a number of promising applications, including quantum computing and communication, quantum key distribution systems, sensing, and metrology [1-3]. Arrays of nonlinear waveguides are one of the most active integrated guided wave optics systems for generating quantum states of light, where quantum protocols can be implemented in miniature scale using integrated optics [4]. The generation and control of nonclassical light in coupled waveguide arrays can be accomplished through various methods, such as the manipulation of several modes [5-7], the number of channels [8, 9], geometry [10], and interactive media such as those used in linear [11-16] and nonlinear processes. Among many others, Kerr effects and second harmonic generation (SHG) can be regarded as the most effective mechanisms in this regard [17-22].

Because the Kerr nonlinear response is incredibly fast and requires no phase matching, it provides an effective mechanism for generating nonclassical effects. Nonlinear waveguides arrays that operate via Kerr effects have produced a variety of fascinating effects, including collapsed and revived mean-photon numbers [23, 24], sub-Poissonian, and photon antibunching [25], squeezing [8, 9], entanglement [22], and quantum steering [21]. Single-mode squeezing has been studied from the perspective of frequency mismatching between the interacting modes. A requirement for achieving an exact and approximative solution for the Hamiltonian has been met [26, 27]. Squeezing was explored in the situation of contra-directional propagation of the input fields [28-30] and the quantum phase features of such devices [31]. The properties for second-order correlation, Bell-like states generation, and chaos have been considered with a train of periodic ultrashort laser pulses [32, 33]. On the experimental side, the squeezing regimes can be accomplished for both pulse and continuous wave cases [34].

Similarly, there are a variety of interaction mechanisms that can be used in the case of SHG. For example, squeezing and entanglement can be achieved in symmetrically coupled waveguides [17, 18]. These can be done assuming codirectional or contradirectional propagation of the input field [26]. Besides that, squeezing can be obtained by integrating asymmetric couplers based on the second-and sub-harmonics generation [35]. Quantum effects can mainly be switched from one channel to another [36]. Quantum Zeno and anti-Zeno effects [37] and higher-order nonclassical effects have all been proposed to account for the symmetrical and asymmetrical coupling of nonlinearities [38]. We recently looked into the quantum properties of multichannel systems with SHG [19, 20]. When adequately controlled parameters are specified, we demonstrated that these systems display a substantially more significant reduction in noise, especially in compound-mode. Furthermore, at least for some interaction regions, the degree of squeezing is less sensitive to the coupling strength between various channels.

While nonlinear systems based on the Kerr effect and SHG are a reliable source of nonclassical light, there is still room for growth, such as an asymmetric multichannel combination of Kerr effects and SHG, which has yet to be developed. As a system with both the second $\chi(2)$ and third-order $\chi(3)$ nonlinearities will emerge with a variety of critical physical dynamics [39], the potentials of increasing the nonclassical effects such as squeezing for interacting Kerr effects and SHG in coupled waveguide arrays are exciting to be addressed. Squeezed light reduces intrinsic quantum noise in field quadrature to below the standard quantum limit. This enables for more exact measurements [40] than would otherwise be achievable and has promising implications for the advancement of large-scale fault-tolerant universal photonic quantum computing [41, 42] and near-thermal quantum advantage via Gaussian boson sampling [43]. The generation of compound-mode squeezing was recently reported utilizing four- and six-wave-mixing techniques [44, 45], and this has potential applications in quantum metrology, communication, and imaging [46].

This chapter aims to investigate the properties of squeezed light in a system of multichannel waveguides with competing $\chi(2)$ and $\chi(3)$ nonlinear effects in terms of quasi-probability distribution functions using normal-ordered phase space representation. We consider a coherently driven single-mode nonlinear waveguide with Kerr effect that is then evanescently coupled with fields from other SHG waveguides. The fundamental harmonic field (FH) is up-converted in pairs to double-frequency second harmonic fields (SH). This lays the groundwork for the development of dense asymmetric multichannel devices.

2. The Model

We consider a nonlinear waveguide with χ(3) nonlinearity in the center, surrounded by f-number of χ(2) nonlinear waveguides, as is shown in Figure 1.

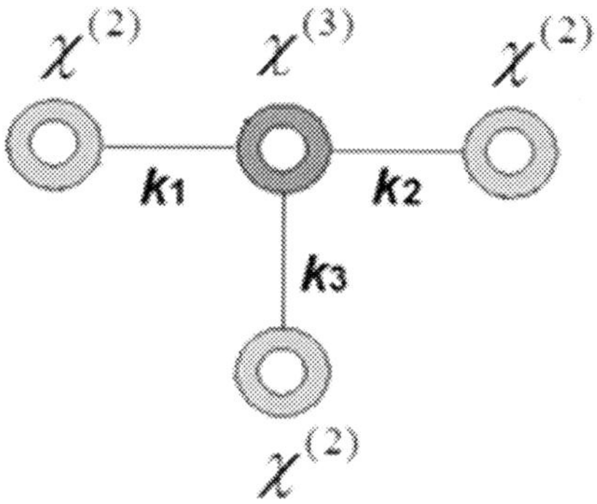

Figure 1. Cross-section of the interaction.

The Hamiltonian of the system is designated by

$$\hat{H} = \hbar\left\{\hat{H}_1 + \hat{H}_2 + \hat{H}_3\right\}, \tag{1}$$

where $\hbar$ is the reduced Planck constant. The first term in the previous Hamiltonian is written as

$$\hat{H}_1 = \omega_1 \hat{a}^\dagger \hat{a} + \omega_2 \sum_{n=1}^{3}\left(\hat{b}_n^\dagger \hat{b}_n\right) + \omega_3 \sum_{n=1}^{3}\left(\hat{B}_n^\dagger \hat{B}_n\right), \tag{2}$$

It depicts the general evolution of the system in a rotating frame. Each waveguide is transversed by the fundamental mode and close enough to permit evanescent coupling. The field in the center channel operating by Kerr effect at frequency ω_1 is evanescently coupled to the fields in the surrounding $\chi^{(2)}$ channels, where an FH field at frequency ω_2 forms an SH field at frequency ω_3. Coefficients $\hat{a}^\dagger \hat{a}$, $\hat{b}_n^\dagger \hat{b}_n$ and $\hat{B}_n^\dagger \hat{B}_n$ $n \in \{1, 2, 3\}$ are bosonic ladder operators satisfying the standard commutation relation $[\hat{O}_i, \hat{O}_j^\dagger = \delta_{ij}$,

$\hat{O}^{\dagger}\hat{O} \Rightarrow \hat{a}^{\dagger}\hat{a}, \hat{b}_n^{\dagger}\hat{b}_n, \hat{B}_n^{\dagger}\hat{B}_n$ for the ω_1, ω_2, and ω_3 fields, respectively. The nonlinear Hamiltonian is denoted by the second term in equation (1) and can be written as

$$\hat{H}_2 = g\hat{a}^{\dagger}\hat{a} + i\frac{\chi}{2}\sum_{n=1}^{3}\left(\hat{b}_n^{\dagger 2}\hat{B}_n - \hat{b}_n^{2}\hat{B}_n^{\dagger}\right). \tag{3}$$

The coefficients g and χ define the magnitude of the self-action Kerr effect and SHG due to the $\chi^{(3)}$ and $\chi^{(2)}$ nonlinear processes in the interactive media. The last term, i.e.,

$$\hat{H}_3 = k_1\left(\hat{a}_1\hat{b}_1^{\dagger} + \hat{b}_1\hat{a}_1^{\dagger}\right) + k_2\left(\hat{a}_1\hat{b}_2^{\dagger} + \hat{b}_2\hat{a}_1^{\dagger}\right) + k_3\left(\hat{a}_1\hat{b}_3^{\dagger} + \hat{b}_3\hat{a}_1^{\dagger}\right), \tag{4}$$

represent the coupling Hamiltonian where k_n, $n \in \{1,2,3\}$ denotes the coupling strength between the $\chi^{(2)}$ and $\chi^{(3)}$ waveguides. The time evolution of the system density operator may be specifically described semi-analytically using the Von-Neumann equation [30]. There are no higher-order derivatives in the Positive P representation. Therefore, it can be precisely mapped into the Fokker-Planck equation using the standard quantum-classical correspondences [47]. Using Ito calculus, the Fokker-Planck equation develops the following set of stochastic differential equations:

$$\dot{\alpha} = -i\left(\omega_1\alpha + 2g\alpha^{*}\alpha^{2} + i\left(k_1\varsigma_1 + k_2\varsigma_2 + k_3\varsigma_3\right)\right) + \sqrt{-2ig}\alpha\eta_1 \tag{5a}$$

$$-\dot{\alpha}^{*} = -i\left(\omega_1\alpha^{*} + 2g\alpha\alpha^{*2} + i\left(k_1\varsigma_1^{*} + k_2\varsigma_2^{*} + k_3\varsigma_3^{*}\right)\right) - \sqrt{2ig}\alpha^{*}\eta_2 \tag{5b}$$

$$\dot{\varsigma}_1 = -i\omega_2\varsigma_1 + \chi\varsigma_1^{*}\xi_1 - ik_1\alpha_1 + \sqrt{\chi\xi_1}\eta_3 \tag{5c}$$

$$\dot{\varsigma}_1^{*} = i\omega_2\varsigma_1^{*} + \chi\xi_1^{*}\varsigma_1 + ik_1\alpha_1^{*} + \sqrt{\chi\xi_1^{*}}\eta_4 \tag{5d}$$

$$\dot{\xi}_1 = -2i\omega_3\xi_1 - \frac{1}{2}\chi\varsigma_1^{2} \tag{5e}$$

$$\dot{\xi}_1^* = 2i\omega_3\xi_1^* - \frac{1}{2}\chi\varsigma_1^{*2} \tag{5f}$$

$$\dot{\varsigma}_2 = -i\omega_2\varsigma_2 + \chi\varsigma_2^*\xi_2 - ik_2\alpha_1 + \sqrt{\chi\xi_2}\eta_5 \tag{5g}$$

$$\dot{\varsigma}_2^* = i\omega_2\varsigma_2^* + \chi\xi_2^*\varsigma_2 + ik_2\alpha_1^* + \sqrt{\chi\xi_2^*}\eta_6 \tag{5h}$$

$$\dot{\xi}_2 = -2i\omega_3\xi_2 - \frac{1}{2}\chi\varsigma_2^2 \tag{5i}$$

$$\dot{\xi}_2^* = 2i\omega_3\xi_2^* - \frac{1}{2}\chi\varsigma_2^{*2} \tag{5j}$$

$$\dot{\varsigma}_3 = -i\omega_2\varsigma_3 + \chi\varsigma_3^*\xi_3 - ik_3\alpha_1 + \sqrt{\chi\xi_3}\eta_7 \tag{5k}$$

$$\dot{\varsigma}_3^* = i\omega_2\varsigma_3^* + \chi\xi_3^*\varsigma_3 + ik_3\alpha_1^* + \sqrt{\chi\xi_3^*}\eta_8 \tag{5l}$$

$$\dot{\xi}_3 = -2i\omega_3\xi_3 - \frac{1}{2}\chi\varsigma_3^2 \tag{5m}$$

$$\dot{\xi}_3^* = 2i\omega_3\xi_3^* - \frac{1}{2}\chi\varsigma_3^{*2} \tag{5n}$$

In the present form of the stochastic differential equation, the over dot represents the system derivation in the direction of z with $\{\varsigma_n, \varsigma_n^*\}$, $\{\xi_n, \xi_n^*\}$, and $\{\alpha, \alpha^*\}$ are the classical representation of independent stochastic fields corresponding to the quantum operator $\hat{a}^\dagger\hat{a}$, $\hat{b}_n^\dagger\hat{b}_n$ and $\hat{B}_n^\dagger\hat{B}_n$ $n \in \{1,2,3\}$ respectively. Parameter η_a $a \in \{1 \text{ to } 8\}$ denotes the Gaussian noise with correlation $\eta_i(z) = 0$ and $\eta_i(z)\eta_j(z') = \delta_{ij}\delta(z - z')$.

To study squeezing, we define the field quadrature on a single-mode basis as:

$$\hat{X}_1 = \frac{1}{2}\left[\hat{O} + \hat{O}^{\dagger}\right]; \quad \hat{Y}_1 = \frac{1}{2i}\left[\hat{O} - \hat{O}^{\dagger}\right]. \tag{6}$$

By substituting $\hat{O}^{\dagger}\hat{O} \Rightarrow \hat{a}^{\dagger}\hat{a}, \hat{b}_n^{\dagger}\hat{b}_n, \hat{B}_n^{\dagger}\hat{B}_n$
all of the individual modes can be obtained. In the form of the stochastic field, i.e., $\hat{O}^{\dagger}\hat{O} \Rightarrow O^{*}O$

$$\left\langle\left(\Delta\hat{X}_1\right)^2\right\rangle = \frac{1}{4}\left\{\left\langle O^2\right\rangle + 2\left\langle O^{*}O\right\rangle + \left\langle O^{*2}\right\rangle + 1 - \left\langle O\right\rangle^2 - 2\left\langle O\right\rangle\left\langle O^{*}\right\rangle - \left\langle O^{*}\right\rangle^2\right\}, \tag{7a}$$

$$\left\langle\left(\Delta\hat{Y}_1\right)^2\right\rangle = \frac{1}{4}\left\{-\left\langle O^2\right\rangle + 2\left\langle O^{*}O\right\rangle - \left\langle O^{*2}\right\rangle + 1 + \left\langle O\right\rangle^2 - 2\left\langle O\right\rangle\left\langle O\right\rangle^{*} + \left\langle O^{*}\right\rangle^2\right\}. \tag{7b}$$

To account for the mixed-mode squeezing, equation (6) can also be written as:

$$\hat{X}_y = \hat{X}_1 + \frac{1}{2}\left[\sum_{j=1}^{l}\left(\hat{P}_j + \hat{P}_j^{\dagger}\right)\right] \quad : \quad \hat{Y}_y = \hat{Y}_1 + \frac{1}{2i}\left[\sum_{j=1}^{n}\left(\hat{P}_j - \hat{P}_j^{\dagger}\right)\right], \tag{8}$$

where $\hat{O}^{\dagger}\hat{O} \Rightarrow \hat{a}^{\dagger}\hat{a}$ and $\hat{P}^{\dagger}\hat{P} \Rightarrow \hat{b}_n^{\dagger}\hat{b}_n, \hat{B}_n^{\dagger}\hat{B}_n$ y, ($y = n + 1$) determines the mode of interaction, i.e., $y = 2$ for two-mode, $y = 3$ for three-mode, and $y = 4$ for four-mode. The maximum mixed-mode interaction, in this case, is limited to $y = 4$, ($f = 3$), and the field quadrature variances are calculated as:

$$\left\langle \left(\Delta\hat{X}_4\right)^2\right\rangle = \frac{1}{4}\left[2\left\langle \sum_{n=1}^{f}\left\{\alpha\varsigma_n + \alpha^*\varsigma_n + \alpha^*\varsigma_n^* + \varsigma_n^*\alpha + \varsigma_1^*\varsigma_n + \varsigma_2^*\varsigma_n + \varsigma_3^*\varsigma_n\right\}\right.\right.$$

$$\left. + \varsigma_1\varsigma_2 + \varsigma_1\varsigma_3 + \varsigma_2\varsigma_3 + \varsigma_1^*\varsigma_2^* + \varsigma_1^*\varsigma_3^* + \varsigma_2^*\varsigma_3^* + \alpha^*\alpha + \frac{1}{2}\left(\alpha\alpha + \alpha^*\alpha^* + \sum_{n=1}^{f}\left\{\varsigma_n\varsigma_n + \varsigma_n^*\varsigma_n^* + f\right\}\right)\right\rangle$$

$$-2\left(\frac{1}{2}\left(\langle\alpha\rangle\langle\alpha\rangle + \langle\alpha^*\rangle\langle\alpha^*\rangle\right) + \sum_{n=1}^{f}\left\{\langle\varsigma_n\rangle\langle\varsigma_n\rangle + \langle\varsigma_n^*\rangle\langle\varsigma_n^*\rangle\right\} + \sum_{n=2}^{f}\left\{\langle\alpha\rangle\langle\varsigma_n\rangle + \langle\alpha^*\rangle\langle\varsigma_n\rangle + \langle\alpha^*\rangle\langle\alpha^*\rangle\right.\right.$$

$$\left. + \langle\varsigma_n^*\rangle\langle\alpha\rangle + \langle\varsigma_1^*\rangle\langle\varsigma_n\rangle + \langle\varsigma_2^*\rangle\langle\varsigma_n\rangle + \langle\varsigma_3^*\rangle\langle\varsigma_n\rangle\right\} + \langle\varsigma_1\rangle\langle\varsigma_2\rangle + \langle\varsigma_1\rangle\langle\varsigma_3\rangle + \langle\varsigma_2\rangle\langle\varsigma_3\rangle$$

$$\left.\left. + \langle\varsigma_1^*\rangle\langle\varsigma_2^*\rangle + \langle\varsigma_1^*\rangle\langle\varsigma_3^*\rangle + \langle\varsigma_2^*\rangle\langle\varsigma_3^*\rangle + \langle\alpha^*\rangle\langle\alpha\rangle\right\}\right)\Big],$$

(9a)

$$\left\langle \left(\Delta\hat{Y}_4\right)^2\right\rangle = \frac{1}{4}\left[-2\left\langle \sum_{n=1}^{f}\left\{\alpha\varsigma_n - \alpha^*\varsigma_n + \alpha^*\varsigma_n^* - \varsigma_n^*\alpha - \varsigma_1^*\varsigma_n - \varsigma_2^*\varsigma_n - \varsigma_3^*\varsigma_n\right\}\right.\right.$$

$$\left. + \varsigma_1\varsigma_2 + \varsigma_1\varsigma_3 + \varsigma_2\varsigma_3 + \varsigma_1^*\varsigma_2^* + \varsigma_1^*\varsigma_3^* + \varsigma_2^*\varsigma_3^* - \alpha^*\alpha + \frac{1}{2}\left(\alpha\alpha + \alpha^*\alpha^* + \sum_{n=1}^{f}\left\{\varsigma_n\varsigma_n + \varsigma_n^*\varsigma_n^* - f\right\}\right)\right\rangle$$

$$-2\left(\frac{1}{2}\left(\langle\alpha\rangle\langle\alpha\rangle + \langle\alpha^*\rangle\langle\alpha^*\rangle\sum_{n=1}^{f}\left\{\langle\varsigma_n\rangle\langle\varsigma_n\rangle + \langle\varsigma_n^*\rangle\langle\varsigma_n^*\rangle\right\}\right) + \sum_{n=1}^{f}\left\{\langle\alpha\rangle\langle\varsigma_n\rangle - \langle\alpha^*\rangle\langle\varsigma_n\rangle + \langle\alpha^*\rangle\langle\alpha^*\rangle\right.\right.$$

$$\left. - \langle\varsigma_n^*\rangle\langle\alpha\rangle - \langle\varsigma_1^*\rangle\langle\varsigma_n\rangle - \langle\varsigma_2^*\rangle\langle\varsigma_n\rangle - \langle\varsigma_3^*\rangle\langle\varsigma_n\rangle\right\} + \langle\varsigma_1\rangle\langle\varsigma_2\rangle + \langle\varsigma_1\rangle\langle\varsigma_3\rangle + \langle\varsigma_2\rangle\langle\varsigma_3\rangle$$

$$\left.\left. + \langle\varsigma_1^*\rangle\langle\varsigma_2^*\rangle + \langle\varsigma_1^*\rangle\langle\varsigma_3^*\rangle + \langle\varsigma_2^*\rangle\langle\varsigma_3^*\rangle + \langle\alpha^*\rangle\langle\alpha\rangle\right\}\right)\Big].$$

(9b)

Squeezing is calculated using the following equations for a single-mode:

$$S_{X,1} = 4\langle(\Delta\hat{X}_1)^2\rangle - |1| \leq 0 \quad : \quad S_{Y,1} = 4\langle(\Delta\hat{Y}_1)^2\rangle - |1| \leq 0. \qquad (10)$$

while for mixed-mode, equation (10) reads as:

$$S_{X,y} = 4\langle(\Delta\hat{X}_y)^2\rangle - |y| \leq 0 \quad : S_{Y,y} = 4\langle(\Delta\hat{Y}_y)^2\rangle - |y| \leq 0. \qquad (11)$$

The system produces squeezed light in the respective quadrature if any field quadrature variances fall below the standard quantum limit, causing the time evolution to be less than 0. The system equation was simulated over 10^3 stochastic trajectories using MATLAB with all model parameters in dimensionless form. The coherent drive in all channels can be depicted using the pseudoprobability distribution $P(O,O^*)=\delta(O-Q)\delta(O^*-Q^*)$, where Q is the coherent state described by $\hat{O}|Q\rangle=Q|Q\rangle$. This means that the same complex conjugate pair Q and Q^* is employed for each stochastic trajectory. This condition is commonly used as an initialization state in quantum optics models, and it is a great way to replicate a well-stabilized laser output [47]. Other input parameters are set at $\omega_1 = 2$, $\omega_2 = 2$, $\omega_3 = 2\omega_2$, $k_1 = k_2 = k_3 = 1$, $g = 10^{-7}$ and $\chi = 10^{-3}$.

3. Discussion

We will now proceed to analyze the properties of squeezed light generated in the system. In Figure 2, we have displayed equation (10) as a function of dimensionless kz. Concerning the Kerr and SHG processes, the field in the $\chi(3)$ (Figure 2(a)) and $\chi(2)$ (Figure 2(b)) waveguides oscillate independently in their respective channels. As the input fields interact with the nonlinear medium, they are no longer in coherent states and are changed to squeeze states for the parameter employed. However, in this scenario, the reduction in fluctuation is due only to the individual nonlinear process in the specific channel. Without evanescent coupling between the channel waveguides, the field quadrature in the $\chi(3)$ channel showed a reduction in fluctuation early in its evolution.

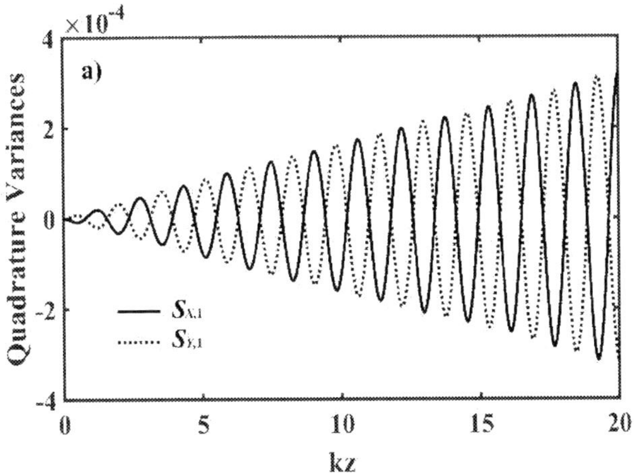

Figure 2. Squeezing without cross-action interaction, $k_1 = k_2 = k_3 = 0$. (a) $\chi^{(3)}$ channel; (b) $\chi^{(2)}$ channel.

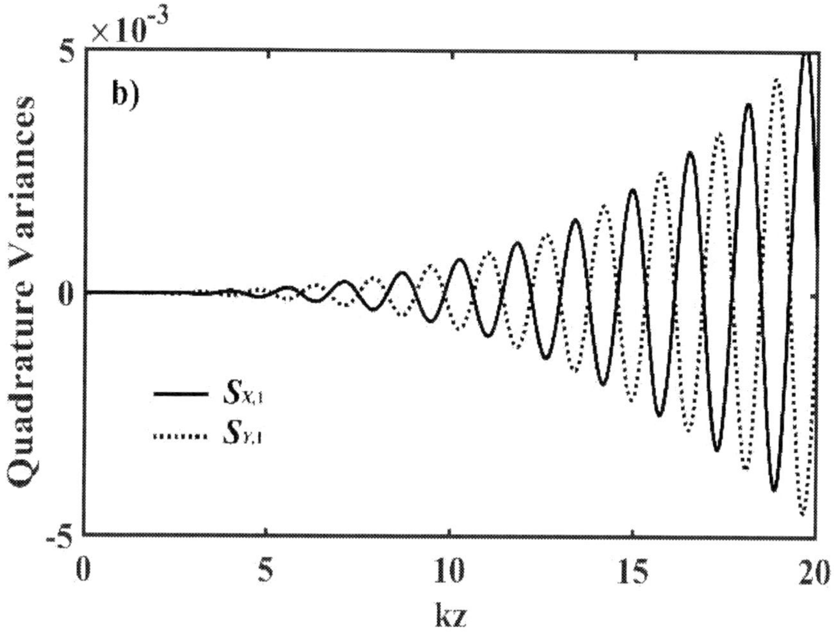

Figure 2. (Continued)

Time development of the quadrature variances indicates a swift oscillation with relatively constant amplitude, which increases as *kz* increases. The evolution of the quadrature fields in the $\chi^{(3)}$ and $\chi^{(2)}$ channels are in phase. Compared to squeezing in the $\chi^{(3)}$ channel, a noticeable increase in the amplitude of squeezing in the $\chi^{(2)}$ channel occurred afterward. Nevertheless, the rate at which squeezing increases as *kz* increases is faster. In general, squeezed light generated by the SHG is stronger than squeezing generated by the Kerr effect. Figure 2 describes the general situation and feature of squeezing in individual channels when the evanescent interaction is absent. We shall show next, the effects of using combined nonlinear interaction to generate squeezing and their attributes.

The results for coupled nonlinear interaction induced by permitting evanescent coupling between the channel waveguide are shown in Figure 3. If the $\chi^{(3)}$ channel interacts with a single $\chi^{(2)}$ waveguide, the period of the quadrature oscillation in the $\chi^{(3)}$ channel is shortened (see Figure 3(a)). As can be observed, the quadrature variance approaches zero, and the phase quadrature consequently surpasses the conventional quantum limit by a much smaller factor. Squeezing in the $\chi^{(3)}$ channel becomes increasingly noticeable as more channel waveguides operating by the SHG process are added to the interaction. For $f = 3$, the mechanism can achieve better squeezing than shown in Figure 2(a). The period of the quadrature oscillation appears to be reducing in the $\chi^{(2)}$ channel, as does the predicted squeezing. Similar to the case observed in the $\chi^{(3)}$ channel, increasing the number of $\chi^{(2)}$ channels may increase squeezing (See Figure 3(b)), although the quantity is weaker than that achieved in Figure 2(b). While the attainment of good squeezing in the $\chi^{(3)}$ channel is significantly dependent on the number of $\chi^{(2)}$ channels, the same cannot be said for the $\chi^{(2)}$ channel. However, the inclusion of $\chi^{(2)}$ channels in the interaction can still be considered reasonable because it permits considerable squeezing. For example, at $f = 3$, the amount of squeezing due to the SHG is equal to that of the Kerr effect.

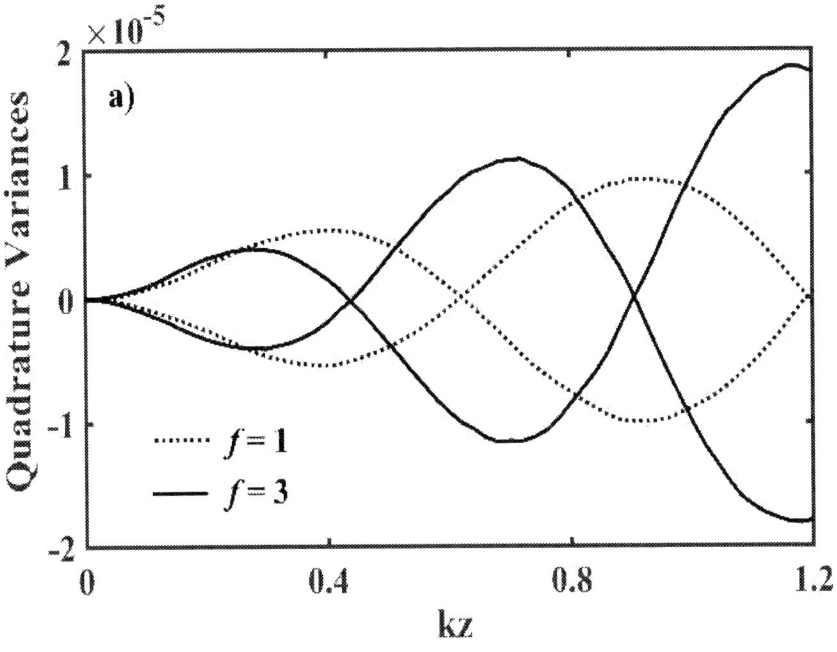

Fgure 3. Squeezing due to the number of $\chi^{(2)}$ channels involved in the cross-action interaction. (a) $\chi^{(3)}$ channel; (b) $\chi^{(2)}$ channel.

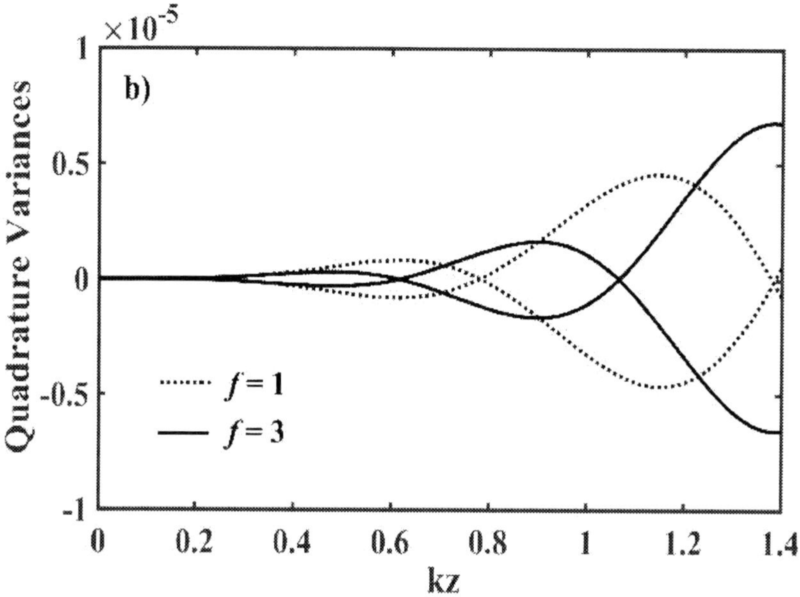

Figure 3. (Continued)

We will look at how frequency mismatch between the $\chi^{(3)}$ and $\chi^{(2)}$ channels affects squeezed light generation. We explore two scenarios in which the system's phase is mismatched when it is first launched. In the first scenario, the frequency of the input field in the $\chi^{(3)}$ channel is higher than that of the $\chi^{(2)}$ channel, resulting in $\Delta_1 = \omega_1\text{-}\omega_2$. In the second scenario, the frequency of the input field in the $\chi^{(3)}$ channel is lower than that of the $\chi^{(2)}$ channel, resulting in $\Delta_2 = \omega_2\text{-}\omega_1$. Figure 4 shows how squeezed light behaves in the $\chi^{(3)}$ and $\chi^{(2)}$ channels at different operational frequencies. Increasing the operating frequency of the coherent field in the $\chi^{(3)}$ channel to $\Delta_1 = 2$ (Figure 4(a)) adds a considerable quantity of noise to the quadrature fluctuation, minimizing squeezing. Squeezing may be recuperated as the frequency is increased further, i.e., $\Delta_1 = 10$. However, the gain is smaller than squeezing attained at $\Delta_1 = 0$. The quadrature variance in the $\chi^{(2)}$ channels exhibits more oscillation with squeezing comparable to that of the $\chi^{(3)}$ channel (Figure 4(b)). Nevertheless, for $\Delta_1 = 10$, squeezing in the $\chi^{(3)}$ channel may surpass squeezing in the $\chi^{(2)}$ channel.

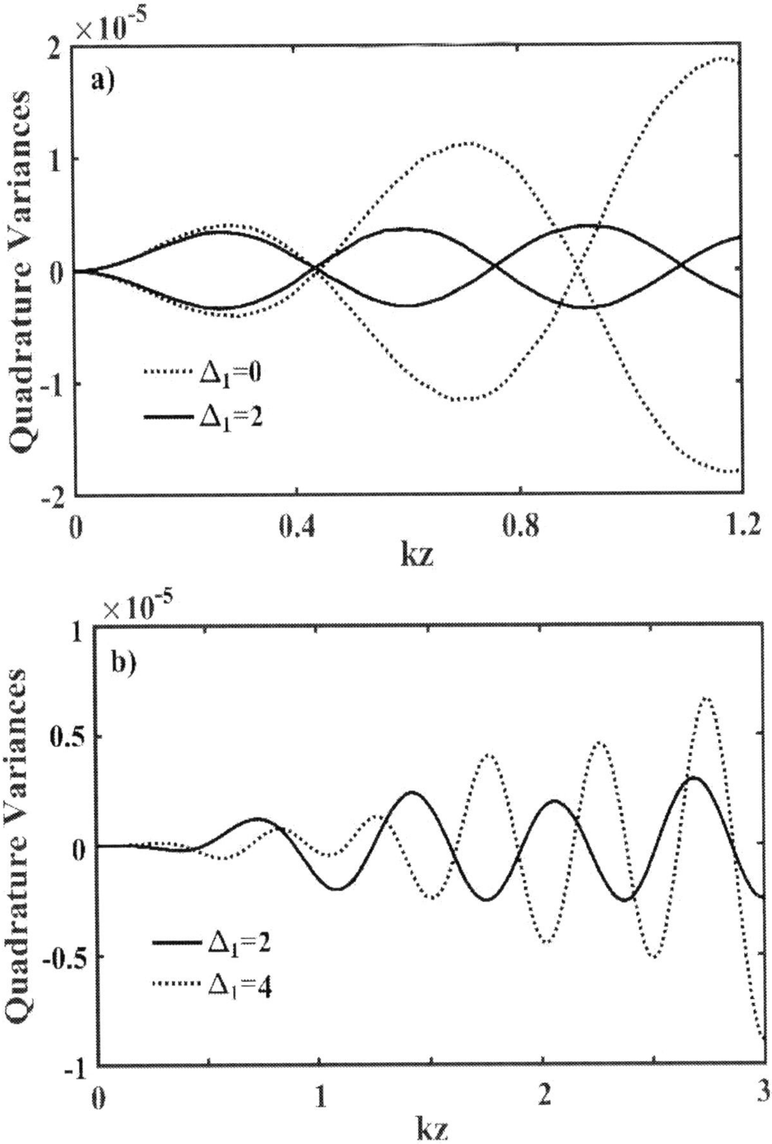

Figure 4. Squeezing with frequencies mismatched $\Delta_1 = \omega_1 - \omega_2$. (a) $\chi^{(3)}$ channel; (b) $\chi^{(2)}$ channel.

While raising the operating frequency of the coherent field in the $\chi^{(3)}$ channel will increase the fluctuation in the quadrature variances, suppression of the quantum noise is maximal if the operating frequencies are increased in the $\chi^{(2)}$ for the second case $\Delta_2 = \omega_2-\omega_1$. Figure 5(a) demonstrates that the fields' quadratures in the $\chi^{(3)}$ channel at $\Delta_1 = 2$ have less noise than $\Delta_2 = 0$. Furthermore, as illustrated in Figure 5(b), raising ω2 will increase squeezing even more. This is, however, constrained to the field at ω1 alone. To improve squeezing in the $\chi^{(2)}$ channel by frequency mismatching, ω_2 must be minimal, for example, $\omega_2 = 2\omega_1$. If ω_2 is significantly larger than ω_1, as seen in Figure 5(c), excess noise will accumulate in the quadrature fields.

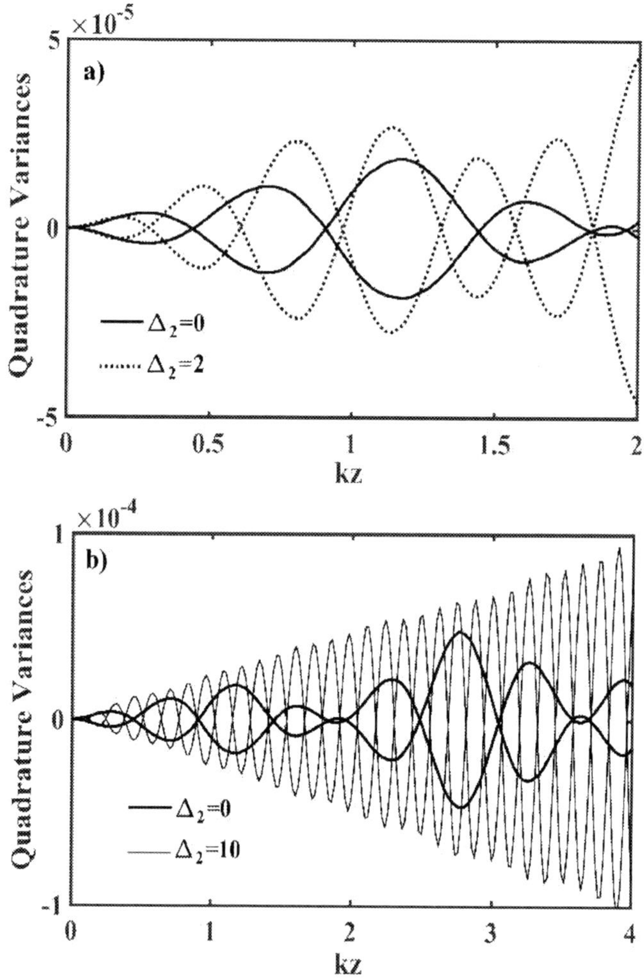

Figure 5. Squeezing with frequencies mismatched $\Delta_2 = \omega_2-\omega_1$. (a), (b) $\chi^{(3)}$ channel; (c) $\chi^{(2)}$ channel.

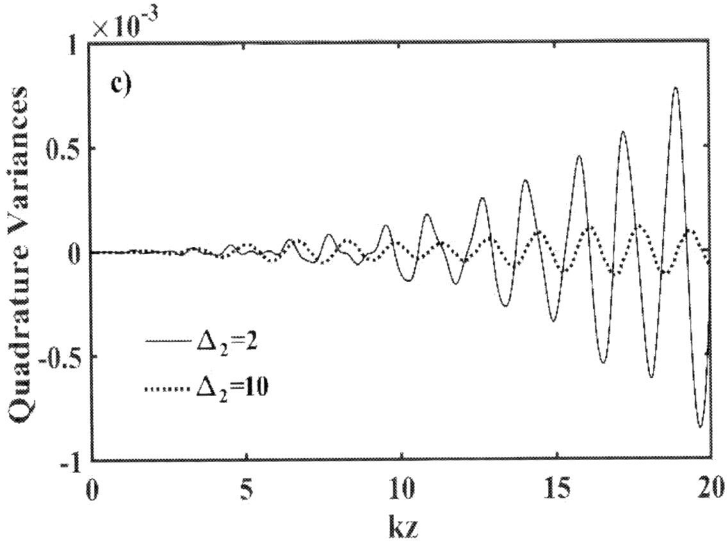

Figure 5. (Continued)

Another advantage of using more channel waveguides in interaction is that it allows for a greater range of quantum correlation options between interaction modes. When only the fields at FH frequency are considered, the correlation between the field modes can be approximated as a four-mode interaction, resulting in compound-mode squeezing on a four-mode basis. Figure 6 demonstrates how the correlation between modes plays a role in the generation of squeezing in the compound-mode basis. The overall effect is that squeezing can be accomplished periodically in the two quadratures in a compound-mode for a large variety of input parameters. Compound-mode squeezing increases as the number of modes increases. A similar occurrence can be observed in the case of frequencies mismatched. For the first case, higher values of $\Delta 1$ will decrease squeezing on a single-mode basis. However, stronger squeezing is observed on a multi-mode basis. Overall, the strongest squeezing in the compound-mode is obtained for the second case of frequency mismatched, i.e., $\Delta 2 = 2$.

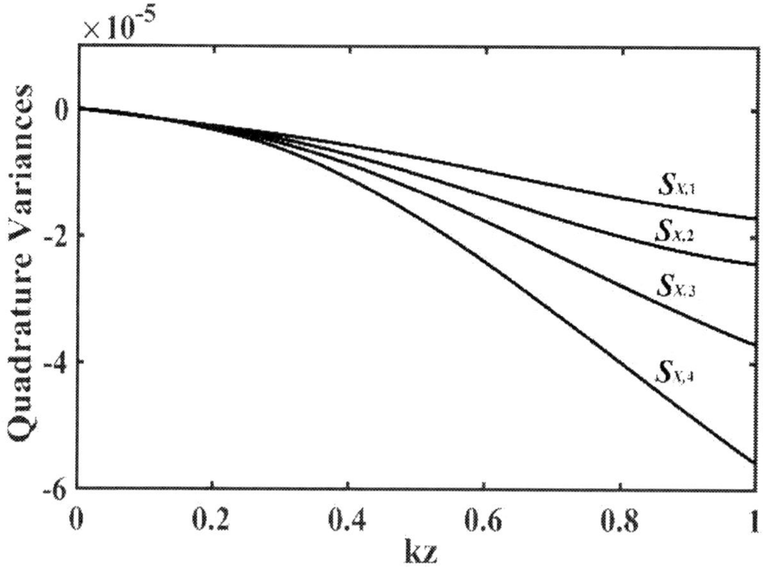

Figure 6. Enhancement of squeezing via compound-mode interaction.

Conclusion

We have studied squeezing properties in an array of asymmetric waveguides induced by the nonlinear Kerr effect and SHG concerning the combined nonlinearities, the number of channel waveguides, frequency mismatched, and compound-mode interaction. Time evolution of the quadrature variances without evanescent coupling between the channel waveguides suggests a fast oscillation that increases as kz expands and squeezed light produced by the SHG is generally more robust than squeezing produced by the Kerr effect. If the waveguides interact, the quadrature variance drop closer to zero and the phase quadrature thus exceeds the standard quantum limit by a smaller factor. However, squeezing increases as the number of $\chi(2)$ channels increases. At f = 3, the amount of squeezing due to the SHG equals that of the Kerr effect. Due to the Kerr effect, squeezing can be improved by frequency mismatched interaction, provided that $\omega 2 > \omega 1$. Other than that, squeezing decreased on a single-mode basis. However, frequency mismatch may result in the generation of stronger squeezing on a multi-mode basis. This shows that coupled waveguide networks with combined nonlinearity can be a valuable source of squeezed light.

Acknowledgments

R Julius gratefully recognizes the Malaysian Ministry of Higher Education (MOHE) national grant FRGS/1/2021/STG07/UITM/02/4.

References

[1] Andersen, U. L. "Photonic chip brings optical quantum computers a step closer," *Nature*, vol. 591, no. 7848, pp. 40–41, 2021.

[2] Tanzilli, S. "Photonic-based quantum communication and sensing: an interplay between fundamental and technological resources," *Conf. on Lasers and Electro-Optics (2020),* p. FF3C.5, 2020.

[3] Takeuchi, S. "Photonic quantum information: science and technology," *Proc. Jpn. Acad. Ser. B*, vol. 92, pp. 29–43, 2016.

[4] Barral D. et al., "Quantum state engineering in arrays of nonlinear waveguides," *Phys. Rev. A*, vol. 102, p. 043706, 2020.

[5] El-Orany F., M. S. Abdalla, and J. Peřina, "Quantum properties of the codirectional three-mode Kerr nonlinear coupler," *Eur. Phys. J. D*, vol. 33, no. 3, pp. 453–463, 2005.

[6] Julius, R. A. -B. M. A. Ibrahim, M. S. M. Deni, and B. Umarov, "Squeezing in four-mode Kerr nonlinear coupler via phase space representation," *AIP Conf. Proc.*, vol. 1557, no. 1, pp. 169–173, 2013.

[7] Julius, R. A. -B. M. A. Ibrahim, and M. S. M. Deni, "Quantum state generation in a four-mode Kerr nonlinear directional coupler," *Laser Phys.*, vol. 24, no. 3, p. 035202, 2014.

[8] Julius, R. A. -B. M. A. Ibrahim, P. K. Choudhury, and H. Eleuch, "On the nonclassical dynamics of cavity-assisted four-channel nonlinear coupler," *Chin. Phys. B,* vol. 27, no. 11, p. 114206, 2018.

[9] Julius, R. A.-B. M. A. Ibrahim, M. K. Abd-Rahman, and P. K. Choudhury, "Quantum dynamics of a four-channel Kerr nonlinear directional coupler," *Opt. Rev.*, vol. 25, no. 5, pp. 563–570, 2018.

[10] Barral D. et al., "Versatile Photonic Entanglement Synthesizer in the Spatial Domain," *Phys. Rev. Applied*, vol. 14, no. 4, p. 044025, 2020.

[11] Aspuru-Guzik A. and P. Walther, "Photonic quantum simulators," *Nat. Phys.*, vol. 8, no. 4, pp. 285–291, 2012.

[12] Tang H. et al., "Experimental two-dimensional quantum walk on a photonic chip," *Sci. Adv.*, vol. 4, no. 5, p. eaat3174, 2018.

[13] Rai, A. G. S. Agarwal, and J. H. H. Perk, "Transport and quantum walk of nonclassical light in coupled waveguides," *Phys. Rev. A*, vol. 78, no. 4, p. 042304, 2008.

[14] Peruzzo A. et al., "Quantum Walks of Correlated Photons," *Science*, vol. 329, no. 5998, pp. 1500–1503, 2010.

[15] Longhi, S. /;"Optical Bloch oscillations and Zener tunneling with nonclassical light." *Phys. Rev. let.,* vol. 101, no. 19, p. 193902. 2008.

[16] Rodríguez-Lara, B. M. "Propagation of nonclassical states of light through one-dimensional photonic lattices," *J. Opt. Soc. Am. B*, vol. 31, no. 4, p. 878, 2014.

[17] Mallon, M. M. Reid, and M. Olsen, "Bright continuous-variable entanglement from the quantum optical dimer," *J. Phys. B At. Mol. Opt. Phys*., vol. 41, no. 1, p. 015501, 2007.

[18] Bache, M. Y. Gaididei, and P. Christiansen, "Nonclassical statistics of intracavity coupled $\chi^{(2)}$ waveguides: The quantum optical dimer," *Phys. Rev. A*, vol. 67, no. 4, p.043802, 2002.

[19] Julius, R. A.-B. M. A. Ibrahim, P. K. Choudhury, and H. Eleuch, "Quantum states generation in multichannel directional coupler with second-order nonlinearity," *Optik,* vol. 186, pp. 212–220, 2019.

[20] Julius, R. A.-B. M. A. Ibrahim, H. Eleuch, and P. K. Choudhury, "Sub-Poissonian photon squeezing and entanglement in optical chain second harmonic generation," *J. Mod. Opt*., vol. 66, no. 10, pp. 1129–1138, 2019.

[21] Midgley, S. L. W. A. J. Ferris, and M. K. Olsen, "Asymmetric Gaussian steering: when Alice and Bob disagree," *Phys. Rev. A*, vol. 81, no. 2, p. 022101,2010.

[22] Leonski W. and A. Miranowicz, "Kerr nonlinear coupler and entanglement," *J. Opt. B Quantum Semiclassical Opt*., vol. 6, no. 3, p. S37, 2004.

[23] Abdalla, M. S. F. A. A. El-Orany, and J. Perina, "Quantum statistics and dynamics of nonlinear couplers with nonlinear exchange," *J. of Mod. Opt.*, vol. 47, no. 6, pp. 1055–1087, 2000.

[24] Ibrahim, A.-B. M. A. B. A. Umarov, and M. R. B. Wahiddin, "Squeezing in the Kerr nonlinear coupler via phase-space representation," *Phys. Rev. A*, vol. 61, no. 4, p. 043804, 2000.

[25] Korolkova, N. and J. Peřina, "Kerr nonlinear coupler with varying linear coupling coefficient," *J. of Mod. Opt.*, vol. 44, no. 8, pp. 1525–1534, 1997.

[26] Peřina, J. "Quantum statistical properties of codirectional and contradirectional nonlinear couplers with phase mismatch," *J. Eur. Opt. Soc. Part B,* vol. 7, no. 5, pp. 863-876, 1995.

[27] Julius, R. M. Zahirzai, A.-B. M. A. Ibrahim, H. Eleuch, and P. K. Choudhury, "Single- and compound-mode squeezing in nonlinear coupler with frequency mismatch," *J. Electromagn. Waves Appl.*, vol. 34, no. 3, pp. 301–315, 2020.

[28] Peřina, J. "Photon statistics of a contradirectional nonlinear coupler," *J. Eur. Opt. Soc. Part B*, vol. 7, no. 5, pp. 849-862, 1995.

[29] Ariunbold, G. and J. Peřina, "Quantum statistics of contradirectional Kerr nonlinear couplers," *Opt. Commun*., vol. 176, no. 1, pp. 149-154, 2000.

[30] Thapliyal, K. A. Pathak, B. Sen and J. Peřina, "Nonclassical properties of a contradirectional nonlinear optical coupler," *Phys. Lett. A*., vol. 378, no. 46, pp. 3431-3440, 2014.

[31] Fiurášek, J. J. Křepelka and J. Peřina, "Quantum-phase properties of the Kerr couplers," *Opt. Commun*., vol. 167, no. 1, pp. 115-124, 1999.

[32] Kalaga, J. M. Jarosik, R. Szczęśniak, T. Nguyen, and W. Leonski, "Pulsed Nonlinear Coupler as an Effective Tool for the Bell-Like States Generation," *Acta Phys. Pol. A*, vol. 135, no.2, pp. 273–275, 2019.

[33] Leonski, W. M. W. Jarosik, R. Szczesniak, J. K. Kalaga, and A. Kowalewska-Kudlaszyk, "Kerr-type nonlinear quantum oscillator: quantum correlations, chaotic, and regular dynamics," in *21st Czech-Polish-Slovak Optical Conference on Wave and Quantum Aspects of Contemporary Optics*, Lednice, Czech Republic, Dec. 2018, p. 4.

[34] Bergman, K. and H. A. Haus, "Squeezing in fibers with optical pulses.," *Opt. Lett.*, vol. 16, no. 9, pp. 663-665, 1991.

[35] Peřina, J. J. Peřina Jr., "Quantum statistics of a nonlinear asymmetric coupler with strong pumping," *J. Eur. Opt. Soc. Part B,* vol. 7, no. 4, pp. 541-552, 1995.

[36] Peřina, J. and *J. Bajer*, "Nonclassical light in nonlinear symmetric and asymmetric couplers," *J. Mod. Opt, vol. 42,* no. 11, p. *2337, 1995.*

[37] Hamilton, C. S. R. Kruse, L. Sansoni, S. Barkhofen, C. Silberhorn, and I. Jex, "Gaussian Boson Sampling," *Phys. Rev. Lett.*, vol. 119, no. 17, p. 170501, 2017.

[38] Thapliyal, K. A. Pathak, B. Sen and J. Peřina, "Higher-order nonclassicalities in a codirectional nonlinear optical coupler: Quantum entanglement, squeezing, and antibunching," *Phys. Rev. A: At. Mol. Opt. Phys.*, vol. 90, no. 1, p. 013808, 2014.

[39] Zhu, E. C. Zhao, and H. Li, "Frequency-domain model of optical frequency-comb generation in optical resonators with second- and third-order nonlinearities," *Phys. Rev. A*, vol. 102, no. 5, p. 053508, 2020.

[40] Schleier-Smith, M. "Squeezing out higher precision," *Science*, vol. 364, no. 6446, pp. 1137–1138, 2019.

[41] Takeda, S. and A. Furusawa, "Perspective: Toward large-scale fault-tolerant universal photonic quantum computing," *APL Photonics*, vol. 4, no. 6, p. 060902, 2019.

[42] Fukui, K. A. Tomita, A. Okamoto, and K. Fujii, "High-Threshold Fault-Tolerant Quantum Computation with Analog Quantum Error Correction," *Phys. Rev. X*, vol. 8, no. 2, p. 021054, 2018.

[43] Hamilton, C. S. R. Kruse, L. Sansoni, S. Barkhofen, C. Silberhorn, and I. Jex, "Gaussian Boson Sampling," *Phys. Rev. Lett.*, vol. 119, no. 17, p. 170501, 2017.

[44] Anderson, B. E. et al., "Phase sensing beyond the standard quantum limit with a truncated SU(1,1) interferometer," *Optica*, vol. 4, no. 7, p. 752, 2017.

[45] Wen, F. et al., "Triple-mode squeezing with dressed six-wave mixing," *Sci. Rep.*, vol. 6, no. 1, p. 25554, 2016.

[46] Rojas-Rojas, S. E. Barriga, C. Muñoz, P. Solano, and C. Hermann-Avigliano, "Manipulation of multi-mode squeezing in a coupled waveguide array," *Phys. Rev. A*, vol. 100, no. 2, p. 023841, 2019.

[47] Olsen, M. K. and A. S. Bradley, "Numerical representation of quantum states in the positive-P and Wigner representations," *Opt. Commun.*, vol. 282, no. 19, pp. 3924–3929, 2009.

About the Editors

Abdel-Baset M. A. Ibrahim received a Ph.D. in Nonlinear Optics (2009) from the University Sains Malaysia (Malaysia). Prior to that, he obtained B.Sc. (in Physics) and M.Sc. (in Quantum Optics) degrees from Al-Zagazig University (Egypt) and the University of Malaya (Malaysia), respectively. During 2000–2005, he was a research officer at Telekom Research and Development (Malaysia) working on optical materials and devices. Since 2010, he has been a Senior Lecturer at the Universiti Teknologi MARA (UiTM; Malaysia), and currently, he is the head of the Photonic Research and Materials group over there. He was a visiting research fellow at Abu Dhabi University (UAE) from September 2018 till December 2020. He has supervised more than 12 M.Sc. and Ph.D. students in the areas of quantum optics, nonlinear optics, and condensed matter physics. He has over 50 publications in international journals and book chapters. He reviews for several photonic-based journals. He is a member of Optica (formerly OSA).

Pankaj K. Choudhury received a Ph.D. degree in physics in 1992. From 1992 to 1997, he was a Research Associate at the Department of Electronics Engineering, Institute of Technology, Banaras Hindu University (Varanasi, India). In 1997, he joined the Department of Physics, Goa University (Goa, India) as a Lecturer. In late 1999, he became a Researcher at the Center of Optics, Photonics and Lasers (COPL), Laval University (Quebec, Canada). From 2000 to 2003, he was with the Faculty of Engineering, Gunma University (Kiryu, Japan), as a

Researcher. In May 2003, he received the position of Professor at the Faculty of Engineering, Multimedia University (Cyberjaya, Malaysia), where he was until late 2009. During that span, he also served the Telekom Research and Development (TMR&D, Malaysia) as a consultant for projects on optical devices. In late 2009, he became Professor at the Institute of Microengineering and Nanoelectronics (IMEN), Universiti Kebangsaan Malaysia (The National University of Malaysia, Malaysia). His research interests lie in the theory of optical waveguides, which include complex mediums, fiber optic devices and nanoengineered structures. He has published over 260 research papers, contributed chapters to 18 books, and edited and co-edited 8 research-level books. He is the reviewer for nearly four dozen research journals. He is the Section Editor of *Optik – International Journal for Light and Electron Optics* (Elsevier, The Netherlands), and the Editor-in-Chief of the *Journal of Electromagnetic Waves and Applications* (Taylor & Francis, UK). He is a Fellow of IET and SPIE, and a Senior Member of the IEEE and Optica (formerly OSA).

List of Contributors

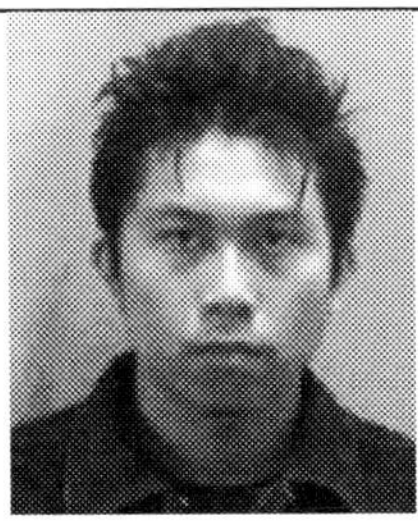

Rafael Julius, PhD
Faculty of Applied Sciences,
Universiti Teknologi Mara (UiTM) Perak,
Tapah Campus, Perak, Malaysia.

Rafael Julius received a B.Sc. degree in Physics (2011), M.Sc., and Ph.D. in Quantum Optics from the Universiti Teknologi MARA (Malaysia) in 2014 and 2019, respectively. He is currently a senior lecturer at the Faculty of Applied Sciences, Universiti Teknologi MARA (Perak).

Abdel-Baset M. A. Ibrahim, PhD
Faculty of Applied Sciences,
Universiti Teknologi MARA (UiTM),
40450 Shah Alam, Selangor, Malaysia
Biography in the "Editors section"

Pankaj K. Choudhury, PhD
Institute of Microengineering and Nanoelectronics,
Universiti Kebangsaan Malaysia,
43600 UKM Bangi, Selangor, Malaysia
Biography in the "Editors section"

Matthew S. Neel, MS
Unaffiliated

Matt earned a BA in Physics from Whitman College and an MS in Physics (emphasis on quantum optics) from Oregon State University. He has worked as a research analyst for the US Department of Energy, a bioinformatics analyst for Fred Hutchinson Cancer Research Center and currently as a senior data analyst for Seattle Children's Research Institute. His research interests include nonlinear optics, magnetic confinement of plasmas and sensor technology. He resides in Seattle, WA, USA.

Mohsen Choubani, PhD
Laboratoire de Micro-Optoélectronique et Nanostructures (LR99ES29),
Département de Physique,
Faculté des Sciences de Monastir 5019,
Avenue de l'environnement,
Université de Monastir, Tunisia

CHOUBANI Mohsen obtained his MPhys in physics from the University of Monastir in 1997 and his Ph.D. in physics from the Scientific Faculty of Tunis in 2011. Since 2010, he has been an accredited Assistant professor in Physics at the Faculty of Science, University of Monastir's Department of Physics. He leads an active research group at the same university that focuses on III–V semiconductor materials and heterostructures (quantum wells, quantum dots), as well as modeling nonlinear optical characteristics of semiconductor nanostructures and photonic applications.

Hassen Maaref, PhD

Laboratoire de Micro-Optoélectronique et Nanostructures (LR99ES29),
Département de Physique,
Faculté des Sciences de Monastir 5019,
Avenue de l'environnement,
Université de Monastir, Tunisia

MAAREF Hassen, received his MPhys in Solid State Physics in 1974 and PhD in Electronic of Semiconductor Materials in 1978 from the University of Paul Sabatier and the National Institute of Applied Sciences at Toulouse in France. Currently, he is a Professor in Semiconductor Physics and Quantum Mechanics in the Department of Physics-Electronic at the Faculty of Sciences of Monastir, University of Monastir since 1979. At the same university, he leads an active research group in the area of epitaxial growth of III–V semiconductor materials and heterostructures (QW, Qdots, HEMT), modelling and characterizations of semiconductor devices for high speed and photonic applications

Faouzi Saidi, PhD
Laboratoire de Micro-Optoélectronique et Nanostructures (LR99ES29),
Département de Physique,
Faculté des Sciences de Monastir 5019,
Avenue de l'environnement,
Université de Monastir, Tunisia

SAIDI Faouzi received his MPhys in 1999 and Ph.D. in Electronic of Semiconductor Materials in 2005 from the University of Tunis and the Scientific Faculty of Tunis in Tunisia. Currently, he is a Professor in Semiconductor Physics and optoelectronic in the Department of Physics at the Institute of Applied Sciences and Technology of Sousse, University of Sousse, in Tunisia since 2014. He leads an active research group in the area of semiconductor materials, heterostructures (QW, Qdots), and characterizations of semiconductor devices for high speed and photovoltaic cells.

Mohd Syafiq M. Hanapi, M.Sc.
Faculty of Applied Sciences,
Universiti Teknologi MARA (UiTM),
40450 Shah Alam, Selangor, Malaysia

Mohd Syafiq M. Hanapi earned his M.Sc. in quantum optics from Universiti Teknologi MARA (UiTM), Malaysia in 2022. Prior to that, he graduated with a B.Sc. (Honours) in Physics and a UiTM- Vice Chancellor's Award in 2019. He also took first place in the Material Lecture Competition (MLC) 2021 in UiTM. He is currently pursuing his Ph.D. in quantum optics at the same institution.

Ahmad M. Alrub, PhD
Department of Physics,
Al-Hussein Bin Talal University,
Ma'an 71111, Jordan

Ahmad Musleh Alrub is currently a Professor of Physics at Al-Hussein Bin Talal University. In 1991, he earned his B.Sc. in physics from the University of Jordan in Amman, and his M.Sc. in nuclear physics from the same institution in 1994. He became a faculty member at Al-Hussein Bin Talal University after receiving his Ph.D. in theoretical condensed matter physics from University Science Malaysia (USM) in 2009. In 2016, he was promoted to Associate Professor, and in 2022, he became Full Professor in the Physics Department. He took a two-year sabbatical at Jouf University in 2018, where he wrote more than sixteen papers on phase transitions. Since 2009, Ahmad has held different administrative posts at Al-Hussein Bin Talal University, including head of department in 2011, deputy dean of academic research and postgraduate studies in 2016, and currently, he is the deputy dean at the faculty of Science.

Index

#

2D material(s), 12

A

algorithm, 52, 54, 55, 56, 57, 84, 85, 87, 89, 90, 98, 173
amplitude, 26, 31, 32, 33, 34, 35, 38, 39, 44, 46, 60, 63, 64, 67, 99, 165, 166, 168, 169, 187, 191, 192, 206
anisotropic, 23, 25, 40, 43, 115
atom(s), 23, 24, 118, 146
autofluorescence, 12

B

beam(s), 8, 19, 43, 47, 48
borate(s), 2, 12

C

calculus, 156, 180, 201
centrosymmetry, 23
characteristic length, 35, 59, 71
characterization, ix, 1, 5, 10, 12, 13, 15, 16
clustering, 3, 11, 12
cluster(s), 9, 11, 12, 13, 15
collagen, 1, 10, 11, 12, 15, 20
communication, 107, 176, 193, 213
computer, 17, 52, 85, 93, 153, 176
computing, 99, 152, 173, 176, 193, 198, 199, 215
conductivity, 86, 95
conductor, 95
conversion efficiency, 11, 12, 68, 69, 73
crystal growth, 1, 10, 11, 12, 15
crystal structure, ix, 10, 11, 13, 23, 24, 45, 46, 50, 51, 65, 78, 81, 111
crystalline, 6, 24, 46, 76, 81
crystal(s), 5, 7, 11, 17, 18, 20, 22, 23, 24, 35, 39, 40, 42, 43, 45, 47, 48, 62, 73, 74, 78, 80, 81, 83, 85, 114, 147

D

database, ix, 1, 3, 5, 15, 55, 56, 57
demultiplexer, 83, 85, 111, 112, 113, 114
density functional theory, 12
depth, 33, 34, 35, 37, 39, 59, 62, 63, 64
derivative(s), 7, 8, 19, 20, 26, 27, 30, 31, 87, 88, 90, 121, 201
dielectric, 7, 8, 11, 12, 16, 19, 45, 47, 48, 81, 83, 86, 99, 101, 104, 109, 111, 150, 152, 174, 195
dielectric permittivity, 86, 109, 111
dielectric(s), 19
differential equation(s), 84, 156, 158, 161, 180, 183, 184, 185, 201
diffraction, 46, 113
dispersion, 7, 19, 39, 81, 84, 85, 97
displacement, 25, 27, 28, 86, 105
distribution function, 199

E

effective index, 40, 66

elastic, 12
electric field, 23, 25, 27, 28, 29, 44, 45, 85, 86, 100, 113, 114, 117, 119, 138, 140, 141, 142, 143, 144, 146, 147, 150, 152
electrical conductivity, 86, 95
electromagnetic, ix, 21, 22, 23, 29, 30, 31, 32, 34, 40, 44, 50, 83, 84, 85, 87, 96, 105, 112, 115, 123
electromagnetic bandgap, 83, 85
electromagnetic wave(s), 29, 30, 31, 32, 34, 44, 84, 115
electron(s), 22, 23, 24, 25, 27, 28, 39, 49, 50, 118, 119, 123, 125, 128, 129, 130, 131, 135, 141, 143
energy, ix, 2, 16, 20, 22, 23, 24, 25, 32, 33, 34, 35, 38, 39, 44, 54, 55, 56, 59, 60, 64, 66, 98, 112, 117, 118, 119, 120, 123, 124, 125, 126, 129, 130, 131, 132, 133, 134, 135, 136, 138, 139, 140, 142, 143, 144, 145, 146, 147, 149, 154, 186, 189
engineering, 4, 24, 78, 84, 198, 213
evolution, 158, 166, 167, 168, 171, 172, 176, 177, 178, 179, 183, 187, 191, 195, 197, 200, 201, 204, 205, 206, 212
excitation, 83, 85, 98, 99, 101, 106, 112, 115
extraordinary index, 40

F

femtosecond laser, 16
fiber(s), 2, 16, 18, 22, 24, 45, 46, 49, 51, 53, 54, 72, 73, 74, 75, 76, 78, 79, 80, 81, 173, 215, 218
fibers, 2, 16, 18, 22, 45, 46, 72, 73, 74, 78, 79, 80, 81, 215
finite element method, 84, 114, 128
fluorescence, 7, 8, 11, 12, 19
frequency conversion, 11, 47
frequency doubling, 6, 11, 21, 32, 49
FDTD, 83
Fresnel coefficient(s), 45, 54, 55
FTIR, 10, 11, 12

G

geometry, 24, 43, 49, 50, 51, 58, 65, 66, 69, 74, 106, 125, 198
growth, 1, 2, 4, 5, 10, 11, 12, 15, 24, 46, 54, 81, 118, 119, 134, 137, 140, 141, 199, 221
growth from solutions, 12
guided mode(s), 54, 64, 65, 66, 67, 68, 69, 70, 71, 72, 74, 76, 77, 78, 79

H

hardness, 12
harmonic generation, 3, 5, 9, 11, 12, 17, 23, 48, 50, 66, 68, 79, 118, 148, 149, 198
Heisenberg equation(s), 182, 183, 184, 185
Heisenberg picture, 153, 158, 177, 178, 192
hyperpolarizability, 12, 20

I

imaging, ix, 10, 11, 12, 199
InAs/GaAs, vii, x, 117, 118, 142, 143, 148, 150
inhomogeneous media, 83, 115
interface, 20, 43, 45, 52, 54, 55, 65, 75, 95, 99, 102
interference, 29, 51, 54, 62, 68, 70, 113
inversion, 23

K

KDP, vii, x, 17, 20, 21, 24, 39, 40, 41, 42, 43, 44, 45, 46, 47, 49, 50, 51, 53, 54, 56, 58, 60, 62, 64, 66, 74, 80, 81
Kerr effect, 198, 199, 200, 201, 206, 207, 212
Kerr nonlinearity, 198
Kleinman symmetry, 29, 31, 43, 44, 47

L

laser(s), ix, 2, 5, 7, 12, 15, 16, 17, 18, 22, 23, 24, 32, 46, 49, 51, 58, 62, 76, 148, 149, 174, 193, 194, 198, 205, 213, 217
lead, 23, 44, 50, 60, 63, 64, 66, 68, 77, 98, 112
lens, 118, 119, 128, 130, 133, 134, 137, 142, 143, 146, 147
light, x, 2, 8, 16, 18, 19, 21, 22, 23, 24, 31, 32, 40, 41, 42, 43, 45, 46, 47, 48, 49, 50, 51, 52, 65, 70, 74, 75, 76, 78, 80, 81, 97, 109, 110, 151, 152, 154, 155, 166, 168, 172, 173, 176, 181, 193, 194, 195, 197, 198, 199, 204, 205, 206, 208, 212, 213, 214, 215
light beam(s), 8, 19, 48
linear optic(s), 22, 23, 47, 152
lithium, 2, 18
lithium niobate, 2, 12, 18
Lorentz model, 24

M

magnetic field(s), 84, 85, 86, 87, 93, 99, 101
magnitude, 6, 26, 28, 37, 54, 63, 73, 133, 135, 137, 140, 141, 144, 147, 180, 201
Maker fringes, 62, 63, 81
Manley-Rowe relations, 34
mass, x, 25, 117, 118, 119, 125, 128
master equation, 153, 155, 179
material(s), ix, 1, 2, 4, 6, 8, 10, 11, 12, 15, 16, 17, 19, 20, 23, 62, 66, 84, 114, 119, 120, 217, 221, 222
matrix, x, 44, 85, 111, 117, 118, 119, 121, 123, 124, 125, 131, 137, 139, 147, 153, 155, 177, 179
matter, iv, x, 22, 40, 144, 217, 223
Maxwell equation(s), 94, 96
microcrystalline, 45, 50, 53, 54, 70, 75, 80
microhardness, 12
microscopy, ix, 7, 8, 11, 12, 19, 20, 128
MOM, 84
multidimensional, 193
multiphoton microscopy, 10, 12, 20

N

nanostructure(s), 2, 7, 8, 17, 19, 83, 117, 118, 148, 220, 221, 222
NATO, 174, 194
negative uniaxial crystal, 39, 41, 43, 44
networking, 15
noncentrosymmetric, 23
nonlinear microscopy, 7, 12
nonlinear optical crystal(s), 12, 39
nonlinear optical effect(s), 5, 39
nonlinear optical material(s), 1, 6, 8, 10, 12, 13, 15, 19, 20
nonlinear optical microscopy, 12
nonlinear optical properties, vii, x, 12, 17, 20, 117, 118, 123, 148, 149
nonlinear optics, ix, 2, 9, 11, 15, 16, 18, 19, 22, 23, 35, 49, 217, 220
nonlinear waveguide(s), 6, 198, 199, 200, 213

O

optical coupler, 151, 152, 173, 174, 175, 198, 214, 215
optical filter(s), 83, 85, 108, 114
optical material(s), 1, 10, 11, 217
optical properties, x, 5, 10, 12, 17, 18, 20, 117, 118, 134, 147
optical studies, 147
ordinary index, 40
organic compound(s), 10, 12, 15
oscillation, 17, 25, 67, 166, 169, 171, 172, 192, 206, 208, 212

P

parallel, 33, 41, 44, 45, 66, 100
perfectly matched layer(s), 83, 85, 94, 115
periodic-poling, 64
permittivity, 86, 95, 101, 103, 109, 111, 125, 152
phase matching, 11, 12, 38, 41, 42, 43, 47, 60, 64, 66, 74, 77, 198

phase mismatch, 33, 34, 37, 38, 41, 54, 59, 60, 62, 63, 65, 67, 68, 69, 71, 73, 77, 190, 191, 192, 195, 214
phase space representation, 194, 198, 199, 213
phase-matching, 6, 35, 37, 41, 44, 58, 60, 62, 63, 64, 66, 67, 68, 70, 75, 76, 77, 79
phase-matching angle, 41, 44, 58, 63, 64, 67
photoluminescence, 12
photonic crystal fiber, 78
photonic crystal(s), 16, 78, 83, 85, 111, 112, 114, 147
photon(s), 2, 22, 32, 176
physics, ii, 4, 21, 25, 55, 80, 217, 221, 223
Planck constant, 154, 200
plasmonic, ii, 2, 12, 16, 17
polarization, ix, 7, 8, 12, 19, 23, 28, 29, 30, 31, 40, 43, 44, 47, 58, 83, 100, 123, 148, 152
potassium, x, 21, 39, 46, 49
potassium dihydrogen phosphate, x, 21, 39, 49
propagation, 7, 19, 40, 56, 63, 74, 76, 80, 97, 102, 107, 153, 154, 172, 175, 176, 177, 178, 179, 181, 182, 183, 192, 198

Q

quantum computer, 153, 176, 213
quantum computing, 152, 173, 176, 198, 199, 215
quantum core/shell, 117, 119, 147
quantum dot, x, 117, 118, 146, 148, 149, 221
quantum mechanics, 22, 153, 155, 158, 179, 192, 194, 195
quantum optics, ix, x, 151, 155, 175, 198, 205, 217, 220, 223
quantum state, 152, 154, 156, 173, 176, 198, 215
quantum well, 148, 149, 221
quasi-phase matching, 11, 12, 64

R

radiation, 7, 17, 18, 24, 30, 50, 54, 64, 65, 78, 107, 112, 113, 149
radiation mode(s), 24, 65, 78
Raman spectroscopy, 6, 12
ray creation and processing algorithm, 54
ray description, 51, 54
reflectivity, 113
refractive index, 97, 106, 124

S

Schrödinger equation, 19, 119, 120, 123, 125, 176, 178
second harmonic generation, iii, vii, viii, ix, 1, 2, 3, 6, 7, 8, 9, 10, 11, 12, 15, 16, 17, 18, 19, 20, 21, 23, 32, 46, 47, 48, 49, 50, 80, 81, 118, 124, 148, 149, 151, 152, 174, 175, 176, 195, 197, 198, 214
second harmonic generation microscopy, 12
second-order nonlinearity, 2, 151, 152, 153, 154, 174, 175, 176, 195, 214
Sellmeier equation(s), 40
semiconductor, 119, 134, 147, 154, 173, 221, 222
sensing, ix, 172, 176, 198, 213, 215
SHG, ix, x, 1, 2, 4, 5, 6, 7, 8, 9, 10, 11, 12, 13, 14, 15, 18, 20, 21, 23, 24, 29, 32, 34, 35, 41, 44, 46, 49, 50, 51, 54, 55, 56, 58, 59, 62, 66, 67, 72, 75, 78, 79, 80, 118, 124, 138, 139, 140, 141, 142, 151, 152, 172, 176, 192, 198, 199, 201, 205, 206, 212
SHG efficiency, 20
silicon, 12, 106, 108, 111, 116
simulation(s), 50, 56, 58, 59, 66, 67, 71, 73, 83, 94, 97, 98, 99, 101, 104, 112, 114, 115, 119, 127, 128, 130, 133, 140, 144, 146, 147, 151, 161, 181, 193, 194
single and coupled quantum dot XE "quantum dot" (s), 117
single crystal, 7, 11, 12, 13, 17, 46, 47
slowly varying amplitude XE "amplitude" approximation, 31

small-signal approximation, 33
solution, 10, 11, 35, 37, 38, 45, 46, 84, 93, 115, 125, 158, 165, 166, 177, 178, 180, 182, 186, 198
solution growth, 11, 12
squeezed state(s), 151, 162, 164, 172, 173, 175, 176, 193, 194, 198
squeezing, x, 151, 152, 154, 162, 164, 165, 166, 168, 169, 170, 171, 172, 173, 174, 175, 186, 187, 188, 189, 190, 191, 192, 193, 194, 195, 197, 198, 199, 202, 203, 204, 205, 206, 207, 208, 209, 210, 211, 212, 213, 214, 215
susceptibility, 29, 43, 152
susceptibility tensor, 29, 43

T

thin film(s), 12, 22, 51, 80
third harmonic generation, 147
total internal reflection, 45, 55, 65, 66, 78, 79
trajectory, 75, 76, 180, 205
transmission, 7, 42, 43, 45, 83, 99, 103, 104, 105, 106, 107, 109, 110, 111, 113, 114, 128, 152
transmission electron microscopy, 128
two-photon excited fluorescence, 12
two-photon microscopy, 12

V

vacuum, 40, 42, 45, 51, 52, 54, 65, 66, 75, 78, 94, 97, 101, 125, 192
vector, 6, 31, 34, 38, 41, 46, 64, 75, 76, 77, 84, 86, 153, 176, 177, 178
velocity, 40, 97, 101
visualization, ix, 1, 3, 9, 13, 164

W

wave description, 51
wave propagation, 40, 80
wave vector, 31, 33, 34, 41, 46, 64, 77
wavelength(s), 23, 24, 43, 50, 51, 110, 113

X

x-ray diffraction (XRD XE "XRD"), 10, 11, 12, 13, 46
XRD, 10, 11, 12, 13, 46

Z

z-scan, 12